NEW COMPUTER METHODS
FOR GLOBAL OPTIMIZATION

MATHEMATICS AND ITS APPLICATIONS

Series Editor: G. M. BELL, Professor of Mathematics,
King's College London (KQC), University of London

NUMERICAL ANALYSIS, STATISTICS AND OPERATIONAL RESEARCH
Editor: B. W. CONOLLY, Professor of Mathematics (Operational Research),
Queen Mary College, University of London

Mathematics and its applications are now awe-inspiring in their scope, variety and depth. Not only is there rapid growth in pure mathematics and its applications to the traditional fields of the physical sciences, engineering and statistics, but new fields of application are emerging in biology, ecology and social organization. The user of mathematics must assimilate subtle new techniques and also learn to handle the great power of the computer efficiently and economically.

The need for clear, concise and authoritative texts is thus greater than ever and our series will endeavour to supply this need. It aims to be comprehensive and yet flexible. Works surveying recent research will introduce new areas and up-to-date mathematical methods. Undergraduate texts on established topics will stimulate student interest by including applications relevant at the present day. The series will also include selected volumes of lecture notes which will enable certain important topics to be presented earlier than would otherwise be possible.

In all these ways it is hoped to render a valuable service to those who learn, teach, develop and use mathematics.

Mathematics and its Applications
Series Editor: G. M. BELL, Professor of Mathematics, King's College London (KQC), University of London

Anderson, I.	Combinatorial Designs
Artmann, B.	The Concept of Number
Arczewski, K. & Pietrucha, J.	Mathematical Modelling in Discrete Mechanical Systems
Arczewski, K. and Pietrucha, J.	Mathematical Modelling in Continuous Mechanical Systems
Bainov, D.D. & Konstantinov, M.	The Averaging Method and its Applications
Baker, A.C. & Porteous, H.L.	Linear Algebra and Differential Equations
Balcerzyk, S. & Joszefiak, T.	Commutative Rings
Balcerzyk, S. & Joszefiak, T.	Noetherian and Krull Rings
Baldock, G.R. & Bridgeman, T.	Mathematical Theory of Wave Motion
Ball, M.A.	Mathematics in the Social and Life Sciences: Theories, Models and Methods
de Barra, G.	Measure Theory and Integration
Bartak, J., Herrmann, L., Lovicar, V. & Vejvoda, D.	Partial Differential Equations of Evolution
Bell, G.M. and Lavis, D.A.	Co-operative Phenomena in Lattice Models, Vols. I & II
Berkshire, F.H.	Mountain and Lee Waves
Berry, J.S., Burghes, D.N., Huntley, I.D., James, D.J.G. & Moscardini, A.O.	Mathematical Modelling Courses
Berry, J.S., Burghes, D.N., Huntley, I.D., James, D.J.G. & Moscardini, A.O.	Mathematical Methodology, Models and Micros
Berry, J.S., Burghes, D.N., Huntley, I.D., James, D.J.G. & Moscardini, A.O.	Teaching and Applying Mathematical Modelling
Blum, W.	Applications and Modelling in Learning and Teaching Mathematics
Brown, R.	Topology
Burghes, D.N. & Borrie, M.	Modelling with Differential Equations
Burghes, D.N. & Downs, A.M.	Modern Introduction to Classical Mechanics and Control
Burghes, D.N. & Graham, A.	Introduction to Control Theory, including Optimal Control
Burghes, D.N., Huntley, I. & McDonald, J.	Applying Mathematics
Burghes, D.N. & Wood, A.D.	Mathematical Models in the Social, Management and Life Sciences
Butkovskiy, A.G.	Green's Functions and Transfer Functions Handbook
Cartwright, M.	Fourier Methods: Applications in Mathematics, Engineering and Science
Cerny, I.	Complex Domain Analysis
Chorlton, F.	Textbook of Dynamics, 2nd Edition
Chorlton, F.	Vector and Tensor Methods
Cohen, D.E.	Computability and Logic
Crapper, G.D.	Introduction to Water Waves
Cross, M. & Moscardini, A.O.	Learning the Art of Mathematical Modelling
Cullen, M.R.	Linear Models in Biology
Dunning-Davies, J.	Mathematical Methods for Mathematicians, Physical Scientists and Engineers
Eason, G., Coles, C.W. & Gettinby, G.	Mathematics and Statistics for the Bio-sciences
El Jai, A. & Pritchard, A.J.	Sensors and Controls in the Analysis of Distributed Systems
Exton, H.	Multiple Hypergeometric Functions and Applications

Series continued at back of book

NEW COMPUTER METHODS FOR GLOBAL OPTIMIZATION

H. RATSCHEK
Professor at the Mathematics Institut der Universität
Düsseldorf, West Germany

and

J. ROKNE
Professor of Computer Science
University of Calgary, Canada

ELLIS HORWOOD LIMITED
Publishers · Chichester

Halsted Press: a division of
JOHN WILEY & SONS
New York · Chichester · Brisbane · Toronto

First published in 1988 by
ELLIS HORWOOD LIMITED
Market Cross House, Cooper Street,
Chichester, West Sussex, PO19 1EB, England
The publisher's colophon is reproduced from James Gillison's drawing of the ancient Market Cross, Chichester.

Distributors:

Australia and New Zealand:
JACARANDA WILEY LIMITED
GPO Box 859, Brisbane, Queensland 4001, Australia

Canada:
JOHN WILEY & SONS CANADA LIMITED
22 Worcester Road, Rexdale, Ontario, Canada

Europe and Africa:
JOHN WILEY & SONS LIMITED
Baffins Lane, Chichester, West Sussex, England

North and South America and the rest of the world:
Halsted Press: a division of
JOHN WILEY & SONS
605 Third Avenue, New York, NY 10158, USA

South-East Asia
JOHN WILEY & SONS (SEA) PTE LIMITED
37 Jalan Pemimpin # 05–04
Block B, Union Industrial Building, Singapore 2057

Indian Subcontinent
WILEY EASTERN LIMITED
4835/24 Ansari Road
Daryaganj, New Delhi 110002, India

© 1988 H. Ratschek and J. Rokne/Ellis Horwood Limited

British Library Cataloguing in Publication Data
Ratschek, H. (Helmut)
Computer methods for global optimization
1. Mathematics. Optimisation. Applications
of computer systems
I. Title   II. Rokne, J, (Jon)
515

Library of Congress CIP available

ISBN 0–7458–0139–0 (Ellis Horwood Limited)
ISBN 0–470–21208–X (Halsted Press)

Typeset in Times by Ellis Horwood Ltd
Printed in Great Britain by Hartnolls, Bodmin

COPYRIGHT NOTICE

All Rights Reserved. No part of this publication may be reproduced, stored in a retrieval system, or transmitted, in any form or by any means, electronic, mechanical, photocopying, recording or otherwise, without the permission of Ellis Horwood Limited, Market Cross House, Cooper Street, Chichester, West Sussex, England.

Contents

	Preface	**1**
1	**Some Principles of Optimization Theory**	**7**
	1.1 Introduction	7
	1.2 Problem Statement	10
	1.3 Optimality Conditions	12
	1.4 Penalty Methods	15
	1.5 Unconstrained Minimization	16
2	**Principles of Interval Analysis**	**23**
	2.1 Introduction	23
	2.2 Why Interval Arithmetic?	25
	2.3 Interval Arithmetic Operations	27
	2.4 Machine Interval Arithmetic	30
	2.5 Further Notations	31
	2.6 Inclusion Functions and Natural Interval Extensions	34
	2.7 Centered Forms, Meanvalue Forms, Taylor Forms	39
	2.8 Improved Interval Hessian Matrices	46
	2.9 Interval Newton Methods	51
	2.10 The Hansen-Greenberg Realization	53
	2.11 Numerical Examples Using the Interval Newton Method	65

3 Global Unconstrained Optimization — 73

- 3.1 Introduction — 73
- 3.2 The Moore-Skelboe Algorithm — 76
- 3.3 Termination, Approximation Errors, Rounding Errors — 81
- 3.4 Convergence Conditions for the Moore-Skelboe Algorithm — 85
- 3.5 Numerical Examples — 89
- 3.6 Convergence Speed of the Moore-Skelboe Algorithm — 95
- 3.7 Convergence Speed with Isotone Inclusion Functions — 102
- 3.8 Ichida-Fujii Algorithm and its Convergence Conditions — 108
- 3.9 Hansen's Algorithm and its Convergence Conditions — 110
- 3.10 Termination Criteria, Approximation Errors and Influence of Rounding Errors — 116
- 3.11 Accelerating Devices: An Overview — 118
- 3.12 Acceleration Devices: Detailed Description — 121
- 3.13 Numerical Examples — 127

4 Unconstrained Optimization over Unbounded Domains — 133

- 4.1 Introduction — 133
- 4.2 The Algorithm over Unbounded Domains — 136
- 4.3 Convergence Properties — 140
- 4.4 The Monotonicity Test — 146
- 4.5 Arithmetic in $\mathbf{I}_\infty$ — 148
- 4.6 Realization on the Computer — 152
- 4.7 Numerical Results — 155

5 Constrained Optimization — 159

- 5.1 Introduction — 159
- 5.2 Problem Statement — 161
- 5.3 Constraints and Exhaustion Principle — 162
- 5.4 Trouble with Constraints — 168
- 5.5 The Basic Algorithm — 172
- 5.6 Optimization Problems with Inexact Data — 177

5.7	Convergence Properties	179
5.8	Accelerating Devices: Overview	183
5.9	Devices for Functions without Differentiability Properties	186
5.10	Devices for $\mathbf{C}^1$ Functions	190
5.11	Devices for $\mathbf{C}^2$ Functions	193
5.12	Numerical Examples	195

Bibliography   199

Notation   223

Index   227

Preface

The *optimization problem* is, in general, to find the optimum (maximum or minimum) value of a function in a given domain and to find the values of the variables where the optimum is reached in this domain. *Global optimization* means to solve the optimization problem in an area given by a real-world problem. *Local optimization* means to solve the optimization problem locally, that is, in the neighborhood of a given point. Local optimization has been investigated in depth; it has a rich theory and many excellent numerical methods and recipes are available. Global optimization, on the other hand, is a recent area which has only been partially researched. Many theories have to be developed and many numerical experiments have to be performed before the area would be considered reasonably well developed. This research and development is, however, of the greatest importance since many real-world problems are global rather than local problems.

If an average user - not too experienced - wants to solve a global optimization problem, he faces two difficulties: The first is that the scope of many books on nonlinear optimization is the development of local methods. To use a local solution as a global solution can - accidentally - lead to a correct result; however, in most cases it will be totally wrong. What shall the user do in order that his local knowledge becomes global? Very few books consider this problem. Indeed too many books suppress the global standpoint because "if one has the local solutions one can easily get the global solution". How can one get "the" local solutions. What happens if the "right" local solution is missing or lost? How can one organize the local information in order to get the global information? To decide which local method shall be chosen in the first place constitutes a further difficulty. There is a beautiful variety of methods discussed in the books, such as linearization methods, penalty and barrier methods, quadratic sequential search methods, methods involving Kuhn-Tucker conditions, clustering and statistical methods, parameter methods, tunnelling and trust region methods, etc. Which method is recommended? Which method is appropriate when searching for the global solution? One certainly cannot give a definite answer since the answer would depend too much on the particular problem of the user: Is the problem large or small, what is known about the problem, how are the differentiability conditions, does he have to solve one problem or many, are then the problems similar or not, has the user plenty of time available or is he busy (wanting the computer to work for him), how is the user's mathematical background, how are the computer costs? Additionally, even the state of art in local optimization has not advanced so far that a generally best method can be given. Thus, the books cannot give a recommendation as to a best method. In spite of this the user has to make a decision if he wishes to solve his problem!

Accordingly, the scope of our book is twofold. The first aim is to *focus on the global problem* and to give precise and computationally reliable instructions on how to organize the relationship between local information and global needs. Some choices of local methods are made which may be combined with the global strategy. The second aim is to *favor interval methods* and to demonstrate that they are excellent tools for handling global problems, i.e., global optimization problems.

Let us make our intentions more clear. First of all, we emphasize a very thorough treatment of the global point of view. The global access we follow is mainly based on the work of Moore, Hansen, Skelboe and of the authors. We give careful convergence theorems of the methods, and we develop the necessary interval analysis background in detail which is necessary in order to understand and apply the global methods. Loosely spoken, a branch and bound principle is used where the bounds can be determined by a computer almost automatically.

Since each global access requires a computationally intensive checking of the whole area (no solution is allowed to be lost), it is necessary to combine the global access with local methods. These local methods can speed up the computation considerably. We will give some examples of how the global access may be combined with local methods depending on the differentiability conditions of the problem (conjugate gradients, Newton methods, etc.) The purpose of these examples is twofold: The more experienced reader can see how interval and non-interval methods are merged and he may then apply knowledge and techniques from his favorite local method as part of the global access. The less experienced reader who does not have the large overview of local methods may use our choices for a complete program. However, we do not discuss the local methods too extensively since they are covered in every monograph on nonlinear optimization. We also do not include linear optimization which is a particular subtopic of its own. Certainly linear programming problems could be solved with our approach, but this would not be very effective.

We have chosen a combination of methods which are characterized to be

- very stable and robust,
- universally applicable,
- 100% reliable,
- flexible and modular,
- convenient,
- applicable without supplying a starting box that usually must contain the set of global minimizers.

The high level of reliability is achieved since safe bounds for the solutions are provided. Further, the algorithms converge to the solutions as far as the assumptions are satisfied. If there is no convergence to the solution then at least inclusions of the solution sets are determined such that the approximation error can be computed easily. It can never happen that the algorithms converge to a wrong result or that they diverge.

Flexibility is provided, as the user may incorporate as much information as he wants to or as he is able to. If he is not willing to provide particular information the program will still reach the solution, but in this case, very slowly.

It is one of the curiosities of global optimization that a starting box must be known which contains the solution even if the problem is so-called unconstrained. We will show how this restriction can be dropped using infinite interval arithmetic.

The number of variables which occur in the optimization problem is practically limited only by storage and computer time. If the number is too large then the problem itself has to be studied first in order to reduce the high computational complexity.

The second aim of this book is to demonstrate that intervals are excellent tools for handling global optimization problems and for supplementing standard techniques. This is because an interval, even though representable by only two points, is an infinite set and is thus a *carrier of an infinite amount of information* which means *global information*. On the contrary, standard optimization methods depend on, and process, local information. For instance, they may evaluate the slope of the objective function at *one* point x, that is $f'(x)$ which is used to approximate also - more or less reliable - the directions of f in some neighborhood of x. Interval arithmetic can however handle expressions like $f'(X)$ for an *interval* X (which can be m-dimensional) which means that $f'(X)$ collects the information $f'(x)$ for any $x \in X$, independent of the size of X.

For example if $0 \notin f'(X)$ - which can be checked automatically during the execution of the program - then one knows that $f'(x) \neq 0$ for all $x \in X$ and that X does not contain a local minimizer (with exemption of the edge). This is global information! Interval arithmetic is therefore particularly suited to dealing with global problems. For example, interval algorithms can determine (enclose) all zeros of a function, or they can determine all solutions of an optimization problem, and that with arbitrary accuracy. This last point has to be emphasized since many people believe that interval methods would lead to unrealistically large intervals.

On the other hand, it can be advantageous to combine interval with non-interval methods, because the latter are frequently local methods and thus faster. For example, if $v = -f'(x)t$ is the direction of steepest descent, then v gives us a local information valid only for x or some neighborhood of x. If instead of v an interval V containing v would be used then V is a direction bundle, and contains clearly the steepest direction v. V will also contain other directions, which may be quite ineffective. As one can see, V collects too much information, which prevents a fast processing such that - in some situations - the computation with the local information, v, will be preferred.

In order to avoid misunderstandings which occur sometimes in connection with the use of interval arithmetic we keep in mind that it is *not the aim* of this monograph to present well-known optimization methods in an interval arithmetical guise in order to control the rounding errors and to generate safe bounds for the errors. The intervals are - as discussed extensively - used as a methodical means for keeping global information. The error control is only a side effect and just sketched in this book.

Necessary background. This monograph is kept on such a mathematical level that readers with a year of calculus, with some basic experience in numerical analysis, and with the intention of thinking in interval dimensions, will not have too much difficulty in reading it.

Abridge of the contents.
Chapter 1 gives a very short introduction to nonlinear optimization. Only a few methods worth combining with interval tools are considered in outline. We prefer standard methods rather than latest developments since the aim of this monograph is to show the applications of intervals to optimization and not to present the state of the art. It is simpler for the reader to understand how the two areas - optimization and intervals - fit together, if techniques that are not too sophisticated are used. The experienced reader, however, who has mastered the merging principles shall not hesitate to apply them to his favorite optimization algorithm if this is possible.

Chapter 2. The interval arithmetical tools are presented as far as they are needed in the main parts of the monograph. Since the computation with intervals and the thinking in intervals are not too widespread, the progress of this chapter is quite gentle, and we tried to make it self-contained. However, we did not include all the proofs which were necessary for a complete treatment.

Chapter 3. Algorithms for the unconstrained optimization problem and their properties are considered. Complete proofs of the convergence theorems are enclosed. The algorithms presented are based on the branch and bound principle. The bounds required are won from an interval arithmetical evaluation of the functions occurring in the problems which implies that the solution data (minimum values, minimum points) is included in boxes at any stage of the algorithms. These algorithms aim to get the including boxes as small as possible or as small as required. Even though the branch and bound principle can be slower than a uniform subdivision method in awkward cases, the interval algorithms compete practically with any other method for solving global problems because of their ability to collect information over arbitrarily large areas.

Chapter 4. The methods and the theory which are developed in Chapter 3 are applied to optimization problems which are unconstrained in the literal sense of the word. This means that the usual assumptions i.e. that the problem is restricted to a compact domain and that the global minimizers are contained - hopefully - in the domain or in its interior, are dropped. Such a generalization is made possible by an appropriate compactification of the real space and by an appropriate realization of infinity arguments on a computer. Complete proofs of the convergence theorems are included.

Chapter 5. The techniques as used for the unconstrained case are mainly provided for the constrained case. The treatment of problems where poor differentiability conditions prevent the application of Lagrangian multiplier methods is emphasized. The algorithms discussed are an interesting combination of interval and non-interval methods. Interval methods are used for exhausting and rejecting unfeasible areas and also for processing global information; non-interval methods are used locally, i.e. for improving the globally obtained information with not too much computational effort. It is, for instance, especially important to find feasible points as fast as possible since the full power of interval methods only comes to play within the feasible domain. This task is also done using interval methods. In contrast, non-interval methods are suggested for finding feasible points taking low function values. The knowledge of such values helps to accelerate the computation considerably. As in the other chapters, a thorough discussion of the convergence properties of the algorithms is included.

Acknowledgements. Thanks are due to the National Sciences and Engineering Research Council of Canada and the Killam Foundation for financial support.

Chapter 1

Some Principles of Optimization Theory

1.1 Introduction

Physicists, chemists, mathematicians, engineers, economists, operations researchers, managers, and practicing computer scientists are often interested in achieving optimal solutions to their problems. These problems may be to determine designs, programs, trajectories, allocation of resources, or approximations of functions. Frequently, different designs or programs, all satisfying the conditions arising from the actual situation, are compared, and one is chosen that also is best in terms of an optimality criterion. Optimization techniques, if properly applied, will automatically examine different designs or plans and select an optimum. Sometimes this is done without solving the complete design or planning problem at every step. But sometimes all possible designs or plans are taken into consideration and a "global" optimum is selected.

A typical and simple problem is the following taken from Simmons (1975):

A chemical company must send 1000 cubic meters of chlorine gas to its research laboratory in another state. Because the gas is extremely dangerous, a special hermetically sealed rectangular railroad car must be built for transporting it. The material from which the top and bottom must be constructed costs $200 per square meter, while the siding material costs half as much; however, only 50 square meters of siding can be obtained. Moreover, the maximum height of the car permitted by tunnels and other overhead clearances is 3 meters. Regardless of the car's dimensions each round trip to the laboratory will cost $800. Assuming no time limit on the overall procedure, what dimensions minimize the total cost of constructing the car and delivering the gas?

Let d, w, and h be the car's length, width, and height. The objective is to minimize overall cost; that is, minimize

$$800(\frac{1000}{dwh}) + 2dw(200) + (2dh + 2wh)(100),$$

where the three terms are contributed by transportation cost, top-and-bottom material, and siding, respectively. The constraints mentioned in the problem are

$$2dh + 2wh \leq 50$$

and

$$h \leq 3.$$

Finally, we must eliminate the possibility of negative dimensions:

$$d, w, h \geq 0.$$

How can such problems be solved in general?

Classical optimization formulas of differential calculus and calculus of variations can be applied to certain optimization problems. There are, however, many problems for which classical formulae are not suitable or may be too cumbersome to apply. For such problems iterative techniques may be appropriate. Optimization problems requiring iterative techniques were practically insurmountable before the advent of modern computers, because of their complexity and the vast amount of computation required. Recently, however, interest in optimization techniques and their application has been considerable, primarily as a result of the developments in computing technology over the last four decades.

Introduction

Let a theoretic, technical or some other real-world system be given. The elements normally involved in the optimization of this system include a system model, an optimality criterion, and an optimization technique. System models of interest here are mathematical models. In order to obtain a mathematical model, the parameters of the system or the independent variables are identified. Then functions of these variables are determined, which represent the different characteristics of the system. The mathematical model is the set of these functions that describes the behavior of the system. The optimality criterion, or objective function, is a measure of merit or cost. It is a function of the same independent variables and yields a number corresponding to each setting of these variables. There exist different optimization techniques, which are applicable to mathematical models and objective functions of different types. It is usually not practical to use the same, very general, optimization technique for all problems.

It is also not the aim of this book to cover all possible optimization techniques available nor to cover some or many of them. In order to get such an overview the reader is referred to the numerous existing excellent textbooks about optimization. It is rather our aim to demonstrate how the tool of interval arithmetic can be used to solve optimization problems. Thus a new class of optimization methods is created. Some of them are independent of the well-known non-interval techniques an ! some of them are dependent on and related to famous formulas such as the Kuhn-Tucker conditions or to popular iterative approaches like Newton methods.

In this chapter a short summary of such classical non-interval formulas and techniques is given as far as they will be combined with interval tools in this text. For an extensive treatment of this classical material the reader is again referred to general textbooks of nonlinear optimization, for example, Bertsekas (1982), Blum-Oettli (1975), Dennis-Schnabel (1983), Evtushenko (1985), Fiacco-McCormick (1968), Fletcher (1980, 1981, 1987b), Gill-Murray-Wright (1981), Hestenes (1975), Himmelblau (1972), Horst (1979), Mangasarian (1969), Mc-Cormick (1983), Minoux (1986), Wolfe, M.A. (1978), Zangwill (1969).

1.2 Problem Statement

In this section we present a precise statement of the problems for which solution techniques are proposed in this book. A general *optimization problem* (sometimes called also a *mathematical or nonlinear programming problem*) can be formulated as follows: Find the values of m variables $x_1, x_2, \ldots, x_m$, denoted for brevity by x, which satisfy the given *constraints*, that is, a given set of equations or inequalities, or both, and optimize (minimize or maximize) the *objective function* $f(x)$. Since the problems of minimizing $f(x)$ and maximizing $-f(x)$ are equivalent, the general optimization problem can be written as:

$$\min_{x \in \mathbf{R}^m} f(x)$$

subject to

$$g_i(x) \leq 0, \quad i = 1, \ldots, k,$$
$$h_i(x) = 0, \quad i = k+1, \ldots, r,$$

where $\mathbf{R}$ denotes the set of real numbers.

In vector notation this general problem is written as

$$\min_{x \in \mathbf{R}^m} f(x) \text{ s.t. } g(x) \leq 0, \; h(x) = 0 \tag{1.1}$$

where $g(x) = (g_1(x), \ldots, g_k(x))^T$ and $h(x) = (h_{k+1}(x), \ldots, h_r(x))^T$. By the superscript "$T$" we mean the transpose of a vector such that we are dealing with column vectors. If there are no constraints ($r = 0$), the problem is said to be *unconstrained*. The constraints may also include as a special case, lower or upper bounds on the variables, that is,

$$a_i \leq x_i \quad \text{or} \quad x_i \leq b_i, \; i = 1, \ldots, m.$$

For practical and computational reasons the unconstrained problem is also assumed to involve lower and upper bounds on the variables. These bounds should be chosen so that the solutions do not occur at the bounds, i.e. it is expected that $a_i < x_i < b_i$ for $i = 1, \ldots, m$. An optimization problem is said to be *linear* if the objective function $f(x)$ and all the constraint functions are linear in the variables

Problem Statement

$x_1, \ldots, x_m$. Otherwise, the optimization problem is said to be *non-linear*. A problem is said to be *quadratic* if the objective function is quadratic,

$$f(x) = a + b^T x + \frac{1}{2} x^T A x$$

where a is a constant, b an m-dimensional column vector, and A an $m \times m$ matrix, and if the constraints are linear (as in the case of a linear problem). If some (or all) of the variables are restricted to a set of integer (or discrete) values, the problem is said to be an *integer* (or a *discrete*) optimization problem.

We assume that the problem statement is formulated such that any set of values of the variables $x = (x_1, \ldots, x_m)^T$ can be interpreted as a point in the m-dimensional column space $\mathbf{R}^m$. By writing down the problem statement in the form (1.1) it is already understood implicitly that the function values $f(x)$, $g_i(x)$ and $h_i(x)$ are real numbers, i.e., using the vector notation, that $g(x) \in \mathbf{R}^k$, $h(x) \in \mathbf{R}^{r-k}$.

A point $x \in \mathbf{R}^m$ satisfying the constraints, that is, $g(x) \le 0$ and $h(x) = 0$ is called a *feasible point*. A set $U \subseteq \mathbf{R}^m$ is called a *feasible set* if all points of U are feasible. The set of all feasible points of problem (1.1) is called *the feasible set* or *the feasible region* of problem (1.1). An inequality constraint $g_i(x) \le 0$ is called *active* if $g_i(x) = 0$, otherwise *inactive* ($i = 1, \ldots, k$). If the problem is unconstrained then there are no constraints and every point is feasible (with respect to the empty set of constraints). However, the attribute feasible is usually dropped in these cases.

A feasible point x^* is called a *local minimum point* or a *local minimizer* for the problem (1.1) if a real number $\epsilon > 0$ exists such that

$$f(x^*) \le f(x) \quad \text{for all feasible points } x$$

with $\| x - x^* \| < \epsilon$. The norm $\|\ \|$ used can be any norm, for example, maximum norm, Euclidean norm, etc. The value $f(x^*)$ is then called a *local minimum* or a *local minimum value*. A feasible point x^* is called a *global minimum point* or a *global minimizer* for the problem (1.1) if

$$f(x^*) \le f(x) \quad \text{for all feasible points } x.$$

The value $f(x^*)$ is then called the *global minimum* or the *global minimum value*. Clearly, a global minimum is a local one, but the converse is generally not true. The concepts *local maximum points, local maximizer*, etc. are defined analogously. By *local optimum points*, etc., we mean either local minimum or maximum points, depending on the context.

Even if one is only interested in methods for solving the global optimization problem (that is, to find global solutions such as global minimizers or global minimum values) one has to pay attention to the local problem. The reason is that several techniques for solving the global problem involve local solutions and local techniques.

1.3 Optimality Conditions

In this section, we summarize some well-known conditions for a point x to be a local minimizer.

The most general optimality condition for problem (1.1) is the *John criterion*. It is essential to distinguish between equality and inequality constraints for this criterion and it is assumed that the objective and constraint functions are differentiable. The criterion says that, if x^* is a local minimizer of (1.1) then vectors $u = (u_0, \ldots, u_k)^T \in \mathbf{R}^{k+1}$ and $v = (v_{k+1}, \ldots, v_r)^T \in \mathbf{R}^{r-k}$ exist such that $x = x^*$ satisfies

$$\left. \begin{array}{l} u_0 f'(x) + \sum_{i=1}^{k} u_i g_i'(x) + \sum_{i=k+1}^{r} v_i h_i'(x) = 0, \\ \sum_{i=1}^{k} u_i g_i(x) = 0, \\ u \geq 0, \\ \begin{pmatrix} u \\ v \end{pmatrix} \neq 0. \end{array} \right\} \quad (1.2)$$

Here

$$f'(x) = (\frac{\partial f(x)}{\partial x_1}, \ldots, \frac{\partial f(x)}{\partial x_m})^T$$

denotes the gradient of f at x which will be identified with the derivative of f at x. Furthermore

$$J_g(x) = (g_1'(x), \ldots, g_k'(x))^T$$

and

$$J_h(x) = (h_{k+1}'(x), \ldots, h_r'(x))^T$$

Optimality Conditions

will denote the Jacobian matrix of g and h at x, respectively. That is

$$J_g(x) = \begin{pmatrix} \dfrac{\partial g_1(x)}{\partial x_1} & \cdots & \dfrac{\partial g_1(x)}{\partial x_m} \\ \vdots & & \vdots \\ \dfrac{\partial g_k(x)}{\partial x_1} & \cdots & \dfrac{\partial g_k(x)}{\partial x_m} \end{pmatrix}$$

and

$$J_h(x) = \begin{pmatrix} \dfrac{\partial h_{k+1}(x)}{\partial x_1} & \cdots & \dfrac{\partial h_{k+1}(x)}{\partial x_m} \\ \vdots & & \vdots \\ \dfrac{\partial h_r(x)}{\partial x_1} & \cdots & \dfrac{\partial h_r(x)}{\partial x_m} \end{pmatrix}.$$

For unconstrained problems, the conditions (1.2) reduce to

$$f'(x) = 0. \tag{1.3}$$

The function whose derivative with respect to x occurs in the first line of (1.2),

$$\Psi(x, u, v) = u_0 f(x) + (u_1, \ldots, u_k) g(x) + v^T h(x),$$

is called the *generalized Lagrangian function* of (1.1).

The necessary conditions of *Kuhn-Tucker* are better known. They require a type of condition called a constraint qualification that guarantees that $u_0 > 0$ in (1.2). Since the known qualifications are rather troublesome to verify in practice they are either more of theoretic interest or only in special cases of importance. It is again assumed that the objective and constraint functions are differentiable. A widely used and well-known version is the following, cf. Horst (1979), Fiacco-McCormick (1968):

Let x be a feasible point of problem (1.1). Let $A(x) = \{i : g_i(x) = 0, \ i = 1, \ldots, k\}$ be the so-called *active index set*. Then x is said to fulfill the *constraint qualification* (also: *regularity condition*) if the set of gradients $g'_i(x), h'_j(x)$ with $i \in A(x)$, $j = k+1, \ldots, m$ is linearly independent.

The Kuhn-Tucker criterion, slightly modified, says: If x^* is a local minimizer of the optimization problem (1.1) and if x^* satisfies the constraint qualification then vectors $u = (u_1, \ldots, u_k)^T \in \mathbf{R}^k$, $v = (v_{k+1}, \ldots, v_r)^T \in \mathbf{R}^{r-k}$ exist such that $x = x^*$ satisfies

$$\left.\begin{array}{l} f'(x) + \sum_{i=1}^k u_i g_i'(x) + \sum_{i=k+1}^r v_i h_i'(x) = 0, \\ \sum_{i=1}^k u_i g_i(x) = 0, \\ u_i \geq 0 \quad \text{for } i = 1, \ldots, k. \end{array}\right\} \quad (1.4)$$

The components u_i, v_i are frequently called the *Lagrangian multipliers*. The function whose derivative with respect to x occurs in the first line of (1.4),

$$L(x, u, v) = f(x) + (u_1, \ldots, u_k) g(x) + v^T h(x)$$

is called the *Lagrangian function* of (1.1).

In order to solve the optimization problem (1.1), the John conditions and Kuhn-Tucker conditions are used as follows: One tries to solve (1.2) or (1.4) for x, u, v with feasible x. If solutions are found then they are checked to see if they are local minimizers of (1.1).

In order to get a sufficient number of equations for x, u, v, conditions (1.2) and (1.4) are completed by the $r - k$ equations

$$h_i(x) = 0, \quad i = k+1, \ldots, r,$$

and by the k equations

$$u_i g_i(x) = 0, \quad i = 1, \ldots, k. \quad (1.5)$$

Equations (1.5) are equivalent to $\sum_{i=1}^k u_i g_i(x) = 0$ for a feasible point x which means that the last-mentioned condition can be deleted. That this condition implies (1.5) can be seen easily: Let us just focus on one fixed i, say j. If j is an active index for x, that is, $g_j(x) = 0$, then $u_j g_j(x) = 0$. If j is an inactive index for x, that is, $g_j(x) < 0$, then $u_j = 0$ follows. Otherwise $u_j g_j(x) < 0$ which would imply $\sum_{i=1}^k u_i g_i(x) < 0$ due to the requirements $u_i \geq 0, g_i(x) \leq 0$.

Thus, in case (1.4), we have $m + r$ equations (plus some restrictions) for the unknowns $x \in \mathbf{R}^m, u \in \mathbf{R}^k, v \in \mathbf{R}^{r-k}$. In case (1.2), we also have $m+r$ equations (plus some restrictions) for the unknowns $x \in \mathbf{R}^m, u \in \mathbf{R}^{k+1}, v \in \mathbf{R}^{r-k}$. One can obtain the missing $(m+r+1)$-th equation by normalizing the multipliers, for example, by

$$\sum_{i=0}^{k} u_i + \sum_{i=k+1}^{r} v_i^2 = 1,$$

cf. also Hansen-Walster (1987b) for further normalizations.

Both systems are now ready to be solved by classical methods such as secant methods, gradient methods, Newton methods, etc.

1.4 Penalty Methods

Penalty methods transform a constrained optimization problem into a sequence of unconstrained optimization problems. The sequence can be finite or infinite depending on the chosen method. The main advantages are that the number of variables does not increase with the transformation, and that the underlying principle is very simple. Disadvantages are that the simpler methods are numerically ill-conditioned, and that the more sophisticated methods lose their simplicity. The reason for this is that these improved versions become effective and avoid the shortcomings of the simpler versions only if they are combined with other theories and techniques, such as duality theory, etc. See, for example, Fletcher (1983, 1985), Bertsekas (1982), Hestenes (1975), and others. Since it is not the scope of this text to present the most perfect way of applying penalty functions but to discuss how to combine interval methods with classical optimization ideas, we only give a brief discussion of exact nondifferentiable penalty functions as a prototype. This prototype may be integrated into more interesting techniques.

The optimization problem (1.1) is again considered. We introduce the function

$$q(x) = \sum_{i=1}^{k} \max(0, g_i(x)) + \sum_{i=k+1}^{r} \mid h_i(x) \mid$$

which takes nonnegative values only. For any x, $q(x) = 0$ iff x is a feasible point.

Thus, the term $q(x)$ together with a so-called *penalty factor* ρ can be used to indicate infeasible points and to push x into the feasible region by increasing ρ. For this purpose we introduce a new objective function,

$$p(x,\rho) = f(x) + \rho q(x), \tag{1.6}$$

which transforms problem (1.1) to the unconstrained optimization problem

$$\min p(x,\rho) \tag{1.7}$$

where $\rho > 0$ is seen as being constant for the moment. $p(x,\rho)$ is frequently called an l_1 *exact penalty function*.

Note that $p(x,\rho)$ need not be differentiable at the boundary of the feasible region. The power of the transformation to (1.7) relies on the fact that if x^* is any local minimizer of (1.1) then a threshold value $\tilde{\rho}$ exists such that x^* is a local minimizer of (1.7) for any $\rho > \tilde{\rho}$. Thus x^* can be computed with a single unconstrained minimization if an appropriate value of ρ is used in (1.7). Otherwise a sequence of problems $\min p(x,\rho_2)$ with $\rho_1 < \rho_2 < \ldots$ must be computed until a local minimizer of $\min p(x,\rho_2)$ is found which is feasible for (1.1). Fortunately, one can obtain good approximations to $\tilde{\rho}$ via estimates of the Lagrangian multipliers.

1.5 Unconstrained Minimization

If problem (1.1) has no constraints, that is, $k = r = 0$, then it is called an *unconstrained minimization problem*, written concisely as

$$\min_{x \in \mathbf{R}^m} f(x) \tag{1.8}$$

where $f : \mathbf{R}^m \to \mathbf{R}$. We use $\mathbf{R}^m$ as domain of f for simplicity. The user can adapt it to his special situation.

If f is differentiable, the classical necessary condition for a point x to be a local minimizer is

$$f'(x) = 0. \tag{1.9}$$

Unconstrained Minimization

A well-known sufficient condition is the following: If (1.9) holds, if f is twice differentiable in x and if

$$f''(x) = \left(\frac{\partial^2 f(x)}{\partial x_i \partial x_j}\right)_{\substack{i=1,\ldots,m \\ j=1,\ldots,m}},$$

the Hessian matrix of f at x, is positive definite then x is a local minimizer.

In connection with interval techniques, we will apply methods for the unconstrained case when we are going to solve

(1) an unconstrained problem by itself,

(2) a constrained problem when the area being processed is already within the feasible region.

Applying techniques for unconstrained minimization can mean two different things:

(i) Looking for a way downhill in order to reach a local minimizer,

(ii) Solving a system of equations, i.e. (1.9), that is the gradient of the objective function f, in order to obtain the critical points of f. The local minimizers are then among the solutions.

Even though these two possibilities look different, they are connected together with the structure of the methods available for the solutions. Good overviews of this subject may be found, for example, in Dennis-Schnabel (1983), Gill-Murray-Wright (1981), Fletcher (1980), McCormick (1983), Ortega-Rheinboldt (1970).

In the remaining part of this section we give a rudimentary sketch of some of the most frequently used methods for treating (i) and (ii).

Newton's Method

Let $\phi : \mathbf{R}^m \to \mathbf{R}^m$ and $\phi \in \mathbf{C}^1$ be given. (By $\mathbf{C}^n$ we mean the class of functions which are n-times continuously differentiable. $\mathbf{C}^1$ functions are also called smooth functions.) We write ϕ but our intent is to mainly consider f', cf. (1.9). The Jacobian matrix of ϕ at x is

denoted by $J(x)$. We want to find a solution $x^* \in \mathbf{R}^m$ of the system of equations
$$\phi(x) = 0. \tag{1.10}$$
The prototype steps of *Newton's method* are:

1. *Choose* $x_0 \in \mathbf{R}^m$.

2. *Set* $n := 0$.

3. *Solve* $J(x_n)s_n = -\phi(x_n)$ *for* $s_n \in \mathbf{R}^m$.

4. *Set* $x_{n+1} := x_n + s_n$.

5. *Set* $n := n + 1$.

6. *Return to 3.*

Advantages are the quadratic convergence property of the method if x_0 is "near" a solution which means that
$$\| x_{n+1} - x^* \| \leq c \| x_n - x^* \|^2$$
for some constant c and any norm of the space $\mathbf{R}^m$. Disadvantages are that each step requires both the computation of $J(x_n)$ and the solution of a system of linear equations.

For practical computations it is frequently recommended to replace the Jacobian matrix of ϕ by finite difference expressions or by secant formulas, cf. Dennis-Schnabel (1983), and others.

A problem which is frequently discussed is the selection of the appropriate strategy when $J(x_n)$ becomes ill-conditioned or singular. There are several ways to overcome this problem. A very promising one, for example, is the tensor method of Schnabel-Frank (1984), cf. also Schnabel (1983). Since the local Newton method is only a tool for speeding up the computation in this monograph, we are not concerned with this problem as long as we use interval methods for solving the global problem (1.1). If the problem occurs then the local search by Newton's method is abandoned and global interval techniques are applied.

If Newton's method is used to solve problem (1.8) directly, that is, solutions of
$$f'(x) = 0$$
are looked for (which are called critical points of f), one has to assume $f \in C^2$ and one obtains *Newton's method for unconstrained optimization* as a special case. The Hessian matrix of f at x is denoted by $f''(x)$. The basic steps are:

1. *Choose $x_0 \in \mathbf{R}^m$.*

2. *Set $n := 0$.*

3. *Solve $f''(x_n)s_n = -f'(x_n)$ for $s_n \in \mathbf{R}^m$.*

4. *Set $x_{n+1} := x_n + s_n$.*

5. *Set $n := n + 1$.*

6. *Return to 3.*

The advantages and disadvantages of Newton's method for unconstrained optimization are the same as for the former Newton's method. There is however a further point to be mentioned, which is that the points x_n tend to a critical point of f. This can be a local minimizer, a local maximizer or a saddle point. Therefore one will apply Newton's method for unconstrained optimization mainly in an area where $f''(x)$ is positive definite which means $z^T f''(x) z > 0$ for $z \in \mathbf{R}^m, z \neq 0$. There are useful adaptations if $f''(x)$ is not positive definite, cf. Gill-Murray-Wright (1981), Dennis-Schnabel (1983), McCormick (1983), Moré-Sorensen (1979), etc. For example, there are good reasons to take modifications of the form
$$f''(x_n) + \mu I$$
instead of $f''(x_n)$ in Step 3 where $\mu > 0$ is chosen such that positive definiteness is obtained. I is the identity matrix.

The use of difference schemes and of secant formulas, as mentioned before, is recommended again in practical computations. Additionally because of the symmetry of $f''(x)$ there are further advantages which lead to interesting developments, such as, for example, the theory of conic models, first proposed by Davidon (1980).

Quasi Newton Methods

As was just mentioned there are many modifications of Newton's method for a variety of reasons. Probably the most important modifications result in a class of algorithms which are "similar" to Newton's method, that is, the class of quasi Newton methods. Unfortunately terminology is neither unique (also: *variable metric methods*; sometimes also: *secant methods*) nor is this class well defined. The first quasi Newton method was suggested by Davidon (1959), cf. Gill-Murray-Wright (1981, p. 125) for further historical details. See also Dennis-Moré (1977).

The key to the success of Newton's method is the curvature information provided by the Hessian matrix which allows a local quadratic model of f to be developed. The theory of quasi Newton methods is based on the fact that an approximation to the curvature information can be obtained without explicitly forming the Hessian matrix, cf. Gill-Murray-Wright (1981). The general iteration step of a *quasi Newton method* for getting a local minimizer of f may be best described by the rule

$$x_{n+1} := x_n - H_n f'(x_n) t_n, \quad n = 0, 1, \ldots$$

where H_n is an approximation to the inverse of the Hessian matrix and t_n is a step-size scalar, cf. McCormick (1983, p. 180). More restrictive are, for instance, Dennis-Schnabel (1983, p. 112), as they use the name quasi Newton method only if the quasi Newton steps are finally replaced by Newton's steps in a sufficiently small neighborhood of the local minimizer.

It is not the aim of this monograph to investigate the theory of quasi Newton methods. Instead we restrict ourselves to mentioning that both the quasi Newton methods and the usual Newton methods may be incorporated into the global interval arithmetical framework as described in later sections.

Unconstrained Minimization

Steepest Descent and Conjugate Gradient Methods

We will summarize here a few methods for solving the unconstrained minimization problem (1.8) where only first order information shall or can be used. Thus we assume $f : \mathbf{R}^m \to \mathbf{R}$ and $f \in \mathbf{C}^1$.

The *steepest descent algorithm (with exact line search)* is so simple in its construction and so widely used that it should at least be mentioned, even though other methods are superior. The basic steps of this algorithm are:

1. *Choose $x_0 \in \mathbf{R}^m$, set $n := 0$.*

2. *If $f'(x_n) = 0$ then stop.*

3. *Set $d_n := -f'(x_n)$.*

4. *Let α_n be a local or global minimizer of the problem $\min f(x_n + \alpha d_n)$ with respect to $\alpha \geq 0$.*

5. *Set $x_{n+1} := x_n + \alpha_n d_n$.*

6. *Set $n := n + 1$. Return to 2.*

It can be proven that the algorithm either terminates at a critical point of f or that the sequence $(f'(x_n))$ converges to zero, provided the set $\{x : f(x) \leq c\}$ is bounded for any constant c. Unless the gradient vanishes, the steepest descent direction, d_n, is clearly a descent direction. However, the number of iterations that are needed to make $\| f'(x_n) \|$ (with any norm) sufficiently small can be very large, even in simple situations such as strictly convex quadratic functions, cf. for example, Powell (1986).

A vast improvement in convergence speed was given by the *conjugate gradient method* of Fletcher-Reeves (1964), although the only change to the above steepest descent steps is Step 3 where the search direction is altered in the following manner:

3. *Set $d_n := \begin{cases} -f'(x_n), & \text{if } n = 0 \\ -f'(x_n) + \beta_n d_{n-1}, & \text{if } n > 0 \end{cases}$*

where $\beta_n = \| f'(x_n) \|^2 / \| f'(x_{n-1}) \|^2$ with the norm being the Euclidean norm.

The *conjugate gradient method of Polak-Ribiére* (1969) is also very successful where β_n for $n > 0$ is defined by

$$\beta_n = f'(x_n)^T(f'(x_n) - f'(x_{n-1}))/ \| f'(x_{n-1}) \|^2 .$$

An interesting comparison of the conjugate gradient methods just described can be found in Powell (1986).

Methods for Nondifferentiable Functions

It would go too far to include methods for nondifferentiable functions to the overview given in this section. We only mention that two cases are mainly distinguished in practice. In the first case, the function still offers some first order information such that methods can be applied which are based on generalized gradients, cf. for example, Demyanov-Vasilev (1985), Zowe (1985), Demyanov-Pallaschke (1985), or the contributions given in Demyanov-Dixon (1986) for a recent overview. In the second case where no appropriate first order information is available, so-called *direct search methods* may be applied. They are based mainly on function value comparisons, cf. for example, Gill-Murray-Wright (1981). See also the methods mentioned in Sec. 3.12.

Chapter 2

Principles of Interval Analysis

2.1 Introduction

Most methods for global optimization suffer from at least one of two defects. The first defect is that the method may not be able to guarantee that the global optimum points have been found to a given tolerance. This means that the results are subject to doubts as regards to their validity. The second defect is that the method may only be applicable under severely restrictive conditions such as the knowledge of a Lipschitz constant, convexity, etc.; otherwise it is likely that only a local minimizer is found instead of a global one. These defects are due to the difficulty of solving the global optimization problem. Global optimization is therefore considered to be an intractable subject.

The above shortcomings may be avoided through the use of the tools and techniques of interval analysis. Within the framework of interval analysis it is possible to develop reliable inclusions for both the minimum values and the minimum points for a very large class of functions.

The work in interval arithmetic began especially with the papers of R.E. Moore, the "father" of interval arithmetic. There are also two very early papers by W. Warmus (1956) from Poland and T. Sunaga

(1958) from Japan who reported on the first investigations in interval arithmetic. Up to now three monographs introducing this subject have appeared. These are R.E. Moore's *Interval Analysis* (1966), also available in German translation (1967), G. Alefeld - J. Herzberger's *Einführung in die Intervallrechnung* (1974), also available in English translation (1983) and R.E. Moore's *Methods and Applications of Interval Analysis* (1979). Many international conferences have also been held where the main theme was interval analysis. The first such conference took place in Oxford, England, in 1968 and the most recent was held in Columbus, U.S.A., in 1987. Proceedings of the symposia have been edited by E. Hansen (1969a), K. Nickel (1975), (1980), (1986), J. Albrycht - H. Wisniewski (1985) and R.E. Moore (1988).

In this chapter some basic tools and techniques of interval analysis are introduced as applied to the global optimization problem. In Sec. 2.2 we motivate and justify the use of interval arithmetic. In Sec. 2.3, the interval arithmetic operations and some basic rules and properties are introduced. In Sec. 2.4, the behavior of interval arithmetic is described when it is executed on a computer. The contents of Sec. 2.5 is mainly concerned with interval arithmetic matrix computations. The main tool for treating optimization problems, that is the concept of an inclusion function, is developed in Sec. 2.6. Natural interval extensions which are also treated in Sec. 2.6 are a constructive means for generating inclusion functions. Sec. 2.7 presents the centered forms, the most popular inclusion functions. We recommend two special centered forms; these are the meanvalue forms and the Taylor forms. These forms can be understood and applied without too much theoretical background. For obtaining optimum bounds of the objective function depending on the information available we give formulas for the developing points which are the best possible chosen for the meanvalue forms. In Sec. 2.8, interval Hessian matrices are introduced that have as few interval entries as possible. Their use in Taylor forms or in interval Newton methods leads to improved approximations. Finally, great care is bestowed upon the presentation of the interval Newton method in Sec. 2.9 and 2.10. Many algorithmic details are added in order to facilitate the implementation of the computational steps of this method.

2.2 Why Interval Arithmetic?

Present-day computers employ an arithmetic called fixed length floating point arithmetic or short, floating point arithmetic. In this arithmetic real numbers are approximated by a subset of the real numbers called the machine representable numbers (or short máchine numbers). Because of this representation two types of errors are generated. The first type of error occurs when a real valued input data item is approximated by a machine number. The second type of error is caused by intermediate results being approximated by machine numbers.

Interval arithmetic provides a tool for estimating and controlling these errors *automatically*. Instead of approximating a real value x by a machine number, the usually unknown real value x is approximated by an interval X having machine number upper and lower boundaries. The interval X contains the value x. The width of this interval may be used as measure for the quality of the approximation. The calculations therefore have to be executed using intervals instead of real numbers and hence the real arithmetic has to be replaced by interval arithmetic. When computing with the usual machine numbers $\tilde{x}$ there is no estimate of the error $\mid \tilde{x} - x \mid$. The computation *with including intervals*, however, provides the following estimate for the absolute error

$$\mid x - \text{mid } X \mid \leq w(X)/2$$

where mid X denotes the midpoint of the interval X and $w(X)$ denotes the width of X. An estimate for the relative error is,

$$\left| \frac{x - \text{mid } X}{x} \right| \leq \frac{w(X)}{2 \min \mid X \mid} \text{ if } 0 \notin X,$$

where $\mid X \mid = \{\mid x \mid : x \in X\}$.

Let us consider an example. The real number 1/3 cannot be represented by a machine number. It may, however, be enclosed in the machine representable interval $A = [0.33, 0.34]$ if we assume that the machine numbers are representable by two digit numbers (without exponent part). If we now want to multiply 1/3 by a real number b which we know lies in $B = [-0.01, 0.02]$ then we seek the smallest interval X which

(a) contains $b/3$,

(b) depends only on the intervals A and B, and does not depend on $1/3$ and b,

(c) has machine numbers as boundaries.

The realization of these requirements is accomplished by two steps,

(i) operations for intervals are defined which satisfy (a) and (b),

(ii) the application of certain rounding procedures to these operations yields (c).

By (i), an interval arithmetic is defined, and by (ii) a machine interval arithmetic is defined.

Let us consider another example where we apply the mean value formula to gain a local approximation of a continuously differentiable function $f : \mathbf{R} \to \mathbf{R}$ ($\mathbf{R}$ denotes the set of reals) near a point $x \in \mathbf{R}$,

$$f(x+h) = f(x) + f'(\xi)h. \tag{2.1}$$

For simplicity, we assume that $h > 0$. Then $\xi \in [x, x+h] =: X$. How can we represent the information given by (2.1) on a computer? How can we evaluate $f(x+h)$ on the computer via the right side of (2.1) if x and h are given? Obviously, ξ is not assigned a numerical value which would be necessary if we wish to compute $f'(\xi)$ automatically on a computer. How should (2.1) be treated in order that it might be used for further numerical manipulation as for example if (2.1) is to be multiplied by a number? The answer is quite simple: Use interval arithmetic and compute

$$F(x, h) := f(x) + f'(X)h$$

as will be defined in the sequel. Then $F(x, h)$ will be an interval, i.e. representable on the computer, and we will know that $f(x+h) \in F(x, h)$ where $f(x+h)$ is unknown and $F(x, h)$ is known.

Such principles have many interesting applications in numerical analysis. Examples are the computational verification of the existence or the uniqueness of solutions of equations in compact domains, cf. Moore (1977, 1978), strategies for finding safe starting regions for iterative methods, cf. Moore-Jones (1977), etc. In this manner one can provide methods which generate a sequence of multi-dimensional intervals, (Y_n), that converges to the set of global minimum points where the sequence $f(Y_n)$ converges to the global minimum of f. A comfortable side effect of the use of interval arithmetic is that when a theoretical interval algorithm is implemented using machine intervals via the so-called outward rounding, the rounding errors are completely under control and cannot falsify the results, cf. Sec. 2.4.

2.3 Interval Arithmetic Operations

Let $\mathbf{I}$ be the set of real compact intervals $[a, b]$, $a, b \in \mathbf{R}$ (these are the ones normally considered). Operations in $\mathbf{I}$ satisfying the requirements (a) and (b) of Sec. 2.2 are then defined by the expression

$$A * B = \{a * b : a \in A, b \in B\} \text{ for } A, B \in \mathbf{I} \qquad (2.2)$$

where the symbol $*$ stands for $+, -, \cdot,$ and $/$, and where, for the moment, A/B is only defined if $0 \notin B$.

The definition (2.2) is motivated by the fact that the intervals A and B include some exact values, α respectively β, of the calculation. The values α and β are generally not known. The only information which is usually available consists of the including intervals A and B, i.e., $\alpha \in A, \beta \in B$. From (2.2) it follows that

$$\alpha * \beta \in A * B \qquad (2.3)$$

which is called the *inclusion principle of interval arithmetic*. This means that the (generally unknown) sum, difference, product, and quotient of the two reals is contained in the (known) sum, difference, product, respectively in the quotient of the including intervals. Moreover, $A * B$ is the *smallest* known set that contains the real number $\alpha * \beta$. Moore (1962) proved that $A * B \in \mathbf{I}$ if $0 \notin B$.

It is emphasized that the real and the corresponding interval operations are denoted by the same symbols. So-called *point intervals*, that is intervals consisting of exactly one point, $[a, a]$, are denoted by a. Expressions like $aA, a + A, A/a, (-1)A$, etc. for $a \in \mathbf{R}, A \in \mathbf{I}$ are therefore defined. The expression $(-1)A$ is written as $-A$.

Definition (2.2) is useless in practical calculations. Moore (1962) proved that (2.2) is equivalent to the following constructive *rules*,

$$\begin{aligned}
{[a,\ b] + [c,\ d]} &= [a + c,\ b + d], \\
{[a,\ b] - [c,\ d]} &= [a - d,\ b - c], \\
{[a,\ b] \cdot [c,\ d]} &= [\min(ac, ad, bc, bd),\ \max(ac, ad, bc, bd)], \\
{[a,\ b] / [c,\ d]} &= [a,\ b] \cdot [1/d,\ 1/c] \text{ if } 0 \notin [c,\ d].
\end{aligned} \quad (2.4)$$

(2.4) shows that *subtraction and division in* $\mathbf{I}$ *are not the inverse operations* of addition and multiplication respectively as is the case in $\mathbf{R}$. For example,

$$\begin{aligned}
{[0,\ 1] - [0,\ 1]} &= [-1,\ 1], \\
{[1,\ 2]/[1,\ 2]} &= [1/2,\ 2].
\end{aligned}$$

This property is one of the main differences between interval arithmetic and real arithmetic. Another main difference is given by the fact that the distributive law of real arithmetic is not valid in general. Only the so-called *subdistributive law*,

$$A(B + C) \subseteq AB + AC \text{ for } A, B, C \in \mathbf{I} \quad (2.5)$$

holds in $\mathbf{I}$. For example,

$$\begin{aligned}
{[0,\ 1][1 - 1]} &= 0, \\
{[0,\ 1]1 + [0,\ 1]1} &= [-1,\ 1].
\end{aligned}$$

The *distributive law* is valid in some special cases, for example,

$$a(B + C) = aB + aC \text{ if } a \in \mathbf{R} \text{ and } B, C \in \mathbf{I}.$$

The following properties follow directly from (2.2): Let $A, B, C, D, \in \mathbf{I}$ and $*$ be any interval operation then

$$\begin{aligned}
&A + B = B + A, \\
&A + (B + C) = (A + B) + C, \\
&AB = BA, \\
&A(BC) = (AB)C, \\
&A \subseteq B, C \subseteq D \text{ implies } A * C \subseteq B * D \text{ (if } B * D \text{ is defined)}.
\end{aligned} \quad (2.6)$$

The last property of (2.6) is the *inclusion isotonicity* of interval operations.

The usual set theoretic operations, *intersection* and *union*, are applicable for intervals $A, B \in \mathbf{I}$, and also for intervals of extended classes as will be introduced later:

$$A \cap B = \{x : x \in A \text{ and } x \in B\},$$
$$A \cup B = \{x : x \in A \text{ or } x \in B\}.$$

Both of these operations can lead to sets not in $\mathbf{I}$. For instance, $A \cap B$ can be empty, and $A \cup B$ can be a set consisting of two disjoint intervals.

A sophisticated extension of the interval arithmetic operations defined above to unbounded intervals is needed and introduced in Chapter 4. For the moment, a small fragment of such an unbounded arithmetic is provided as far as it is required for the interval Newton method in Sec. 2.9. Alefeld (1968) was the first to use infinite intervals in Newton methods. The following formulas are due to Hansen (1980):

Let $0 \in [c, d]$ and $c < d$, then

$$[a, b]/[c, d] = \begin{cases} [b/c, +\infty) & \text{if } b \leq 0 \text{ and } d = 0, \\ (-\infty, b/d] \cup [b/c, +\infty) & \text{if } b \leq 0, c < 0, \text{ and } d > 0, \\ (-\infty, b/d] & \text{if } b \leq 0 \text{ and } c = 0, \\ (-\infty, a/c] & \text{if } a \geq 0 \text{ and } d = 0, \\ (-\infty, a/c] \cup [a/d, +\infty) & \text{if } a > 0, c < 0, \text{ and } d > 0, \\ [a/d, +\infty) & \text{if } a \geq 0 \text{ and } c = 0, \\ (-\infty, +\infty) & \text{if } a < 0 \text{ and } b > 0, \end{cases} \quad (2.7)$$

and furthermore $[a, b]/0 = (-\infty, \infty)$.

These formulas are not applicable to every problem, but they are appropriate for solving linear equations in connection with the interval Newton method. There is also no need for implementing the formulas (2.7) on the machine explicitly since they are finally intersected with a bounded interval such that the result is always a bounded interval, a pair of bounded intervals, or the empty set. We also have to shift unbounded intervals before intersecting them. This means that for

$x, a \in \mathbf{R}$,
$$\begin{aligned} x + [a, \infty) &= [x+a, \infty), \\ x + (-\infty, \infty) &= (-\infty, \infty), \\ x + (-\infty, a] &= (-\infty, x+a]. \end{aligned}$$

It is hardly possible to deal with larger interval arithmetic calculations unless formulas and rules are available for common combinations of intervals so that it is not necessary to resort to the fundamental arithmetic rules (2.4) for each calculation. For a good collection of such formulas and their proofs the reader is referred to Alefeld-Herzberger (1983). Examples of such formulas are, where $w([c, d]) = d - c$ and mid $[c, d] = (c+d)/2$:

$$\begin{aligned} w(aA \pm bB) &= |a| w(A) + |b| w(B), \\ \mathrm{mid}(aA \pm bB) &= a \,\mathrm{mid}\, A \pm b \,\mathrm{mid}\, B \end{aligned}$$

for $a, b \in \mathbf{R}$, $A, B \in \mathbf{I}$. If A is *symmetric*, that is, $A = [-a, a]$ for some $a \geq 0$, and if $B = [c, d]$ then

$$\left.\begin{aligned} AB &= A\max(|c|, |d|), \\ A/B &= \begin{cases} A/c \text{ if } c > 0, \\ A/d \text{ if } d < 0, \end{cases} \\ w(AB) &= 2a \max(|c|, |d|), \end{aligned}\right\} \quad (2.8)$$

etc.

2.4 Machine Interval Arithmetic

Let us return to the requirements (c) or (ii) of Sec. 2.2, that is, that the endpoints of our intervals must be machine numbers. This leads to a special topic called *machine interval arithmetic*. It can be considered as an approximation to interval arithmetic on computer systems.

Machine interval arithmetic is based on the inclusion isotonicity of the interval operations in the following manner: Let us again assume that α, β are the unknown exact values at any stage of the calculation, and that only including intervals are known, $\alpha \in A$, $\beta \in B$. Then A, B might not be representable on the machine. Therefore A and B are replaced by the *smallest machine intervals* that contain A and B,

$$A \subseteq A_M, \quad B \subseteq B_M.$$

A machine interval is an interval which has left and right endpoints that are machine numbers. From (2.6) it follows that

$$A * B \subseteq A_M * B_M.$$

The interval $A_M * B_M$ need not be a machine interval and it is therefore approximated by $(A_M * B_M)_M$ which is representable on the machine. This leads to the *inclusion principle of machine interval arithmetic*:

$$\alpha \in A, \beta \in B \text{ implies } \alpha * \beta \in (A_M * B_M)_M. \qquad (2.9)$$

Thus, the basic principle of interval arithmetic is retained in machine interval arithmetic, that is, the exact unknown result is contained in the corresponding known interval, and *rounding errors are under control*.

We *sum up*: When a concrete problem has to be solved then our procedure is as follows:

- the theory is done in interval arithmetic;
- the calculation is done in machine interval arithmetic;
- the inclusion principle (2.9) provides the transition from interval arithmetic to machine interval arithmetic.

There are several software systems and software packages in which machine interval arithmetic is implemented, for example TRIPLEX-ALGOL-60, PASCAL-SC, FORTRAN-SC, or ACRITH for some IBM computers, ARITHMOS for some Siemens computers, etc.

2.5 Further Notations

The set of real numbers is denoted by **R** and the set of real compact intervals $[a, b], a, b \in \mathbf{R}$ by **I**. If $A \in \mathbf{I}$ then we also write $A = [\text{lb}A, \text{ub}A]$ denoting the lower and upper boundaries of A by $\text{lb}A$ and $\text{ub}A$. Intervals of the form $[a, a]$ are called *point intervals*. These are identified with the corresponding reals. The *width* of an interval $A = [a, b]$ is denoted by

$$w(A) = b - a$$

and the *midpoint* by
$$\text{mid } A = \frac{a+b}{2}.$$

If $A = [a,\ b], B = [c,\ d] \in \mathbf{I}$ then
$$A \leq B \text{ iff } b \leq c$$
and
$$A < B \text{ iff } b < c.$$

If $D \subseteq \mathbf{R}$ then $\mathbf{I}(D) = \{Y \mid Y \in \mathbf{I}, Y \subseteq D\}$.

The set of real m-dimensional column vectors is denoted by $\mathbf{R}^m$ and the set of m-dimensional interval column vectors by $\mathbf{I}^m$. The superscript T means the transpose of a row or a column.

If $A = (A_1, \ldots, A_m)^T \in \mathbf{I}^m$ then A is a right parallelepiped $A_1 \times A_2 \times \ldots \times A_m$. The vector of left endpoints of A is denoted by $\text{lb}A = (\text{lb}A_1, \ldots, \text{lb}A_m)^T$ and the vector of right endpoints is denoted by $\text{ub}A = (\text{ub}A_1, \ldots, \text{ub}A_m)^T$. Interval vectors are also called intervals when it is clear from the context whether real intervals or interval vectors are intended. A *box* is also a frequently used synonym for an m-dimensional interval vector in this monograph.

If $A = (A_1, \ldots, A_m)^T \in \mathbf{I}^m$ then the *width* of A is defined to be
$$w(A) = \max\ \{w(A_i) : i = 1, \ldots, m\}$$
and the *midpoint* of A to be
$$\text{mid } A = (\text{mid } A_1, \ldots, \text{mid } A_m)^T.$$

The set of $n \times m$ real matrices is denoted by $\mathbf{R}^{n \times m}$ and the set of $n \times m$ interval matrices by $\mathbf{I}^{n \times m}$. If $A = (A_{ij}) \in \mathbf{I}^{n \times m}$ then
$$\text{mid } A = (\text{mid } A_{ij}) \in \mathbf{R}^{n \times m}$$
is the *midpoint* of A. Clearly mid $A \in A$.

If $A, B \in \mathbf{I}, A, B \in \mathbf{I}^m$ or $A, B \in \mathbf{I}^{n \times m}$ then
$$A \cap B = \{x : x \in A \text{ and } x \in B\}$$

is the *intersection* of A and B and

$$A \cup B = \{x : x \in A \text{ or } x \in B\}$$

the *union* of A and B.

If $A = [a,\ b], B = [c,\ d] \in \mathbf{I}$ then the *inclusion* of A in B is as usual

$$A \subseteq B \text{ if } c \leq a \leq b \leq d.$$

If $A, B \in \mathbf{I}^m$ or $A, B \in \mathbf{I}^{n \times m}$ then $A \subseteq B$ means that $A_i \subseteq B_i$ for $i = 1, \ldots, m$ or $A_{ij} \subseteq B_{ij}$ for $i = 1, \ldots, n; \ j = 1, \ldots, m$. Similarly, if $x \in \mathbf{R}^m$, $A \in \mathbf{I}^m$, or if $x \in \mathbf{R}^{n \times m}, A \in \mathbf{I}^{n \times m}$ then

$$x \in A$$

means $x_i \in A_i$ for $i = 1, \ldots, m$ or $x_{ij} \in A_{ij}$ for $i = 1, \ldots, n; \ j = 1, \ldots, m$.

Similarly, if $A, B \in \mathbf{I}^m$ then

$$A \leq B \quad \text{or} \quad A < B$$

means $A_i \leq B_i$ for $i = 1, \ldots, m$, or $A_i < B_i$ for $i = 1, \ldots, m$, respectively. Note that $A \leq B$ does *not* mean that $A = B$ or $A < B$ holds as is the case with inequalities in $\mathbf{R}$.

The interval arithmetic operations are extended to *interval vector* and *interval matrix operations* in the usual manner:

$$\begin{aligned}
a(A_{ij}) &= (aA_{ij}) \text{ for } a \in \mathbf{R},\ (A_{ij}) \in \mathbf{I}^{n \times m} \\
(A_{ij}) \pm (B_{ij}) &= (A_{ij} \pm B_{ij}) \text{ for } (A_{ij}), (B_{ij}) \in \mathbf{I}^{n \times m} \\
(A_{ij})(B_{ij}) &= (\sum_{l=1}^{k} A_{il} B_{lj}) \text{ for } (A_{ij}) \in \mathbf{I}^{n \times k},\ (B_{ij}) \in \mathbf{I}^{k \times m}.
\end{aligned}$$

This definition includes the arithmetic for interval vectors (rows as well as columns) by setting $n = 1$ or $m = 1$.

If A, B, C, D are interval vectors or interval matrices and if $*$ denotes one of the operations $+, -, \cdot,$ or $/$, then

$$A \subseteq C,\ B \subseteq D \text{ implies } A * B \subseteq C * D \tag{2.10}$$

if the operations are defined. Property (2.10) is the extended form of the *inclusion isotonicity* of the interval arithmetic operations.

Letting $x, y \in \mathbf{R}$, we define (after Sunaga (1958))

$$x \vee y = \begin{cases} [x, y] & \text{if } x \leq y, \\ [y, x] & \text{if } y \leq x. \end{cases}$$

We see that $x \vee y$ is the interval spanned by the reals x and y, that is, the smallest interval that contains x and y. If $x = (x_1, \ldots, x_m)^T$, $y = (y_1, \ldots, y_m)^T \in \mathbf{R}^m$, then we define

$$x \vee y = (x_1 \vee y_1, \ldots, x_m \vee y_m)^T \in \mathbf{I}^m.$$

Clearly, $x \vee y$ is the interval that is spanned by the vectors x and y. If $X, Y \in \mathbf{I}$ then $X \vee Y$ denotes the smallest interval which contains X and Y. If $X = [a, b]$ and $Y = [c, d]$ then

$$X \vee Y = [\min(a, c), \max(b, d)].$$

Furthermore, if $X = (X_1, \ldots, X_m)^T, Y = (Y_1, \ldots, Y_m)^T \in \mathbf{I}^m$ then the smallest interval $Z \in \mathbf{I}^m$ that contains X and Y is denoted by $X \vee Y$ and it is equal to $(X_1 \vee Y_1, \ldots, X_m \vee Y_m)^T$.

For $D \subseteq \mathbf{R}^m$ we denote by $\mathbf{I}(D)$ the set of all boxes $Y \in \mathbf{I}^m$ with $Y \subseteq D$. For example, if $X \in \mathbf{I}^m$, and thus $X \subseteq \mathbf{R}^m$, the set of all subboxes Y of X is just $\mathbf{I}(X)$. In this connection we say that Y is an *interval variable* over $\mathbf{I}(X)$ which shall mean that Y can take each box of $\mathbf{I}(X)$ as value. This terminology is mainly used when functions $F : \mathbf{I}(X) \to \mathbf{I}$ etc. are considered.

2.6 Inclusion Functions and Natural Interval Extensions

In this section, the inclusion principle for the interval arithmetic operations is extended to more general functions such as programmable functions.

In the treatment of optimization problems using interval arithmetic the main tool is the concept and application of inclusion functions. Let $D \subseteq \mathbf{R}^m$ and $f : D \to \mathbf{R}$. Let furthermore $\Box f(Y) = \{f(x) :$

$x \in Y\}$ be the *range* of f over $Y \in \mathbf{I}(X)$. A function $F : \mathbf{I}(D) \to \mathbf{I}$ is called an *inclusion function* for f if

$$\Box f(Y) \subseteq F(Y) \quad \text{for any } Y \in \mathbf{I}(X). \tag{2.11}$$

Inclusion functions for vector-valued or matrix-valued functions are defined analogously. The inclusion condition (2.11) must in this case be satisfied componentwise.

It turns out that interval analysis provides a natural framework for constructing inclusion functions recursively for a large class of functions.

In order to outline this class of functions it is assumed that some fundamental functions are available for which inclusion functions are already known. This assumption is verified by existing computer languages for interval computations. These languages have pre-declared functions g (examples are sin, cos, etc.) available. For these functions it is also assumed that *pre-declared inclusion functions* G satisfying the above conditions (2.11) are given. The functions G are easy to construct since their monotonicity intervals are generally known, such that even $G(Y) = \Box g(Y)$ will hold, in general. It is also easy to realize these inclusion functions G on a computer such that (2.11) is not violated. In this case the influence of rounding errors is kept under control by computing

$$(G(Y_M))_M \text{ instead of } G(Y),$$

cf. Sec. 2.4.

Let $f : D \to \mathbf{R}$, $D \subseteq \mathbf{R}^m$ be a function which may be described in some programming language as an explicit expression without use of logical or conditional statements (such as "if ... then", "while", etc.). That is, each function value $f(x), X \in D$, can be written down as an expression (also denoted by $f(x)$) which is independent of the value of x and such that this expression consists only of

(1) the variable x (or its components $x_1, \ldots, x_m$),

(2) real numbers (coefficients, constants),

(3) the four arithmetic operations in $\mathbf{R}$,

(4) the pre-declared functions g,

(5) auxiliary symbols (parentheses, brackets, commas, etc.).

Let $Y \in \mathbf{I}(D)$ then the *natural interval extension* of f to Y is defined as that expression which is obtained from the expression $f(x)$ by replacing each occurrence of the variable x by the box Y, the arithmetic operations of $\mathbf{R}$ by the corresponding interval arithmetic operations, and each occurrence of a pre-declared function g by the corresponding inclusion function G. This definition is due to Moore (1966). The natural interval extension of $f(x)$ to Y is denoted by $f(Y)$ when defined as an expression, that is, a string of some specified symbols. The function value which is obtained by evaluating this expression is also denoted by $f(Y)$.

It follows from the inclusion isotonicity of the interval operations, (2.6), and from the properties of the pre-declared inclusions, i.e. the G's, to be inclusion functions (see (2.11)) that

$$x \in Y \text{ implies } f(x) \in f(Y). \qquad (2.12)$$

Since property (2.12) is the key to almost all interval arithmetic applications and results, E. Hansen suggested at an international interval workshop in Columbus, Ohio, September 1987, that (2.12) should be called the *fundamental property of interval arithmetic*. We have just followed this suggestion.

If $f : D \to \mathbf{R}, D \subseteq \mathbf{R}^m$ is programmable and can be described by a function expression as characterized above then the interval function $F : \mathbf{I}(D) \to \mathbf{I}$ defined by $F(Y) = f(Y)$ is an inclusion function for f. More importantly we have here an effective constructive means to find an inclusion function F for a real function f using the tool of natural interval extensions.

(Note: Natural interval extensions could only be precisely defined via a recursion. Further, one would have to distinguish between the expressions $f(x)$ or $f(Y)$ and the functions defined by these expressions as mentioned briefly. One would also have to take care of the case that a forbidden division by 0 could be implied by evaluating $f(Y)$ even if $0 \notin Y \in \mathbf{I}(D)$. Although such distinctions are mathematically unavoidable we prefer omitting them since they would cause confusion rather than clarity and help. The reader preferring a more precise presentation is referred to Ratschek-Rokne (1984).

Example. If $f(x) = x_1 \sin(x_2) - x_3$ for $x \in \mathbf{R}^3$ and if SIN is the pre-declared interval function for sin then $f(Y) = Y_1 \text{ SIN}(Y_2) - Y_3$ is the natural interval extension of f to $Y \in \mathbf{I}^3$.

It is one of the large curiosities of interval arithmetic that different expressions for one and the same function f lead to interval expressions which are also different as functions:

Example. If $f_1(x) = x - x^2$ and $f_2(x) = x(1-x)$ then $f_1(x)$ and $f_2(x)$ are different as expressions, but equal as functions. Further, $f_1(Y)$ and $f_2(Y)$ are also different as functions, i.e., if $Y = [0, 1]$ then $f_1(Y) = Y - Y^2 = [-1, 1]$, $f_2(Y) = Y(1-Y) = [0, 1]$. For comparison, $\Box f(Y) = [0, 1/4]$.

It is therefore a very important and challenging problem to find expressions for a given function that lead to the best possible natural interval extensions, that is, $f(Y)$ shall approximate $\Box f(Y)$ as well as possible. Part of the solution to this problem can be found in Ratschek-Rokne (1984).

Up to now, we have only admitted programmable functions not containing logical connectives for the application of the natural interval extension. The reason is that the existence of such connectives can cause the inclusion isotonicity to fail. This is best shown by an example.

Example. The function $f : \mathbf{R} \to \mathbf{R}$ shall be defined by

$$f(x) = \begin{cases} x, & \text{if } x \geq 0, \\ 0, & \text{otherwise.} \end{cases}$$

By automatically applying the principle of natural interval extension, that is, by replacing x by intervals $Y \in \mathbf{I}$, one would get the following interval function,

$$f(Y) = \begin{cases} Y, & \text{if } Y \geq 0, \\ 0, & \text{otherwise.} \end{cases}$$

If $Y = [-1, 1]$ then $f(Y) = 0$ but $\square f(Y) = [0, 1]$, and thus, the function $F : \mathbf{I} \to \mathbf{I}$, defined by $F(Y) = f(Y)$ for $Y \in \mathbf{I}$, is no longer an inclusion function of f. After a moment's reflection it is easily seen that an appropriate inclusion function may also be found in this case (and also in other cases). This inclusion function $F_1 : \mathbf{I} \to \mathbf{I}$ is defined by

$$F_1(Y) = \begin{cases} Y & \text{if } Y \geq 0, \\ 0 & \text{if } Y \leq 0, \\ [0, \text{ub}Y], & \text{otherwise.} \end{cases}$$

This example shows that it is almost always possible to find an inclusion function for a programmable function. (The restriction "almost" means that we did not try to prove it. On the other hand, we do not know a counterexample.) It also shows that inclusion functions for general programmable functions cannot be constructed as automatically as in the case of functions having explicit expressions.

A measure of the quality of an inclusion function F of f is the *excess-width*,

$$w(f(Y)) - w(\square f(Y)) \quad \text{for } Y \in \mathbf{I}(D),$$

introduced by Moore (1966). A measure for the asymptotic decrease of the excess-width as $w(Y)$ decreases is the so-called order (also: convergence order) of F, due to Moore (1966): An inclusion function F of $f : D \to \mathbf{R}, D \subseteq \mathbf{R}^m$ is called of *(convergence) order* $\alpha > 0$ if

$$w(F(Y)) - w(\square f(Y)) = \mathcal{O}(w(Y)^\alpha),$$

that is, if there exists a constant $c \geq 0$ such that

$$w(F(Y)) - w(\square f(Y)) \leq c w(Y)^\alpha \quad \text{for } Y \in \mathbf{I}(D).$$

In order to obtain fast computational results it is important to choose inclusion functions of an order α as high as possible when $w(Y)$ becomes small. A detailed investigation of the order of inclusion functions is given in Ratschek-Rokne (1984). A similarly looking concept, which is however independent of the order, is the idea of a Lipschitz function. Let $D \subseteq \mathbf{R}^m$ and $F : \mathbf{I}(D) \to \mathbf{I}^k$. Then F is called *Lipschitz* if there exists a real number K *(Lipschitz constant)* such that

$$w(F(Y)) \leq K w(Y) \quad \text{for } Y \in \mathbf{I}(D).$$

The Lipschitz property delivers us a frequently used criterion for the meanvalue form which is a special inclusion function being of convergence order 2, cf. the next section.

2.7 Centered Forms, Meanvalue Forms, Taylor Forms

Centered forms are inclusion functions with special features that were introduced by Moore (1966). Many important contributions have been made to the development of centered forms including a general definition. It is not possible to present these contributions here; however, the interested reader is referred to the monograph by Ratschek-Rokne (1984). This monograph gives an extensive overview starting from Moore's (1966) first historical, rather vague, definition proceeding with the investigation by Krawczyk-Nickel (1982) resulting in an elegant axiomatic characterization, and ending with the higher order forms of Cornelius-Lohner (1984).

In order to construct an inclusion function for a given function, one has two choices in general, i.e. one can choose between natural interval extensions and centered forms. The most important centered forms are the meanvalue form and the Taylor form, both of which are described in this section.

Let $f : D \to \mathbf{R}, D \subseteq \mathbf{R}^m$ be differentiable and let $F' : \mathbf{I}(D) \to \mathbf{I}^m$ be an inclusion function for the gradient, f'. Then $T_1 : \mathbf{I}(D) \to \mathbf{I}$ defined by

$$T_1(Y) = f(c) + (Y - c)^T F'(Y) \quad \text{for } Y \in \mathbf{I}(D) \qquad (2.13)$$

where $c = \text{mid } Y$ or also some other point of Y is called the *meanvalue form function* (or shorter: *meanvalue form*) of f, cf. Moore (1966), (1979). Frequently, $F'(Y)$ can be computed via natural interval extensions or via an automatic differentiation arithmetic, cf. Rall (1981), or via similar techniques that avoid explicit differentiation. Because of the meanvalue formula of analysis we have, if $Y \in \mathbf{I}(D)$ is given, for $x \in Y$,

$$f(x) = f(c) + (x - c)^T f'(\xi) \in f(c) + (Y - c)^T F'(Y)$$

where ξ is a point between x and c. It is thus obvious that the meanvalue form is an inclusion function for f. Its importance arises from its second order property which is obtained with a low computational effort:

THEOREM 1 *(Krawczyk-Nickel (1982)).* *If F' is Lipschitz then the meanvalue form T_1 is of convergence order 2.* □

An extensive proof of this theorem can be found in Ratschek-Rokne (1984).

Example. Let $f(x) = x - x^2$ be defined on $D = \{x : x \geq 1\} \subseteq \mathbf{R}$. ($D$ instead of $\mathbf{R}$ is chosen for simplicity in order to avoid different cases.) An inclusion function for $f'(x) = 1 - 2x$ is

$$F'(Y) = 1 - 2Y \quad \text{for } Y \in \mathbf{I}(D)$$

(natural interval extension of f'). The meanvalue form of f is

$$T_1(Y) = (c - c^2) + (y - c)(1 - 2Y) \quad \text{for } Y \in \mathbf{I}(D)$$

where $c = \text{mid } Y$. The natural interval extension of $f(x)$ to Y is

$$f(Y) = Y - Y^2 \quad \text{for } Y \in \mathbf{I}(D).$$

Finally,

$$\Box f(Y) = [x^2 - x,\ y^2 - y] \quad \text{for } Y = [x, y] \in \mathbf{I}(D).$$

Let us look at the widths of the inclusion functions:

$$\begin{aligned} w(\Box f(Y)) &= y^2 - y - (x^2 - x) = y^2 - x^2 - (y - x) \\ &= w(Y)(y + x - 1). \end{aligned}$$

Using (2.8) and the fact that

$$\max\{|\,1 - 2x\,|,\ |\,1 - 2y\,|\} = 2y - 1$$

for $Y = [x, y] \in \mathbf{I}(D)$ we get

$$\begin{aligned} w(T_1(Y)) &= w[(Y - c)(1 - 2Y)] \\ &= w(Y - c)(2y - 1) \\ &= w(Y)(2y - 1). \end{aligned}$$

The width of the natural interval extension is
$$\begin{aligned} w(f(Y)) &= (y-x)+(y^2-x^2) \\ &= w(Y)(y+x+1). \end{aligned}$$
A short calculation shows that
$$w(T_1(Y)) \le w(f(Y)) \text{ iff } w(Y) \le 2,$$
which means that the meanvalue form is superior for smaller intervals Y. This is consistent with the fact that the meanvalue form is of convergence order 2, but the natural interval extension is only of order 1:
$$\begin{aligned} w(T_1(Y)) - w(\Box f(Y)) &= w(Y)^2 = \mathcal{O}(w(Y)^2), \\ w(f(Y)) - w(\Box f(Y)) &= w(Y)^2 = \mathcal{O}(w(Y)). \end{aligned}$$

From this example it is clear that it is not always wise to take a meanvalue form - especially for boxes Y with larger width - since its excess-width tends quadratically to ∞ as $w(Y) \to \infty$ whereas the excess-width of the natural interval extension tends only linearly to ∞. This situation is typical for the whole area of inclusion functions such that meanvalue forms as well as Taylor forms should only be used if $w(Y) \le 1/2$. This is an average recommendation and results from our own numerical experience.

Remarks. 1. One obtains, in general, meanvalue forms with smaller widths if *slopes* instead of $F'(Y)$ are used in (2.13). The interested reader is referred to Krawczyk (1983), Alefeld-Herzberger (1974), Krawczyk-Neumaier (1985), Ratschek-Rokne (1984).

2. The quality of the chosen centered form, for instance the meanvalue form, depends on the shape of the function being included such that for special functions special centered forms are superior, cf. for example Alefeld-Rokne (1981), Rokne (1986).

Let $f : D \to \mathbf{R}, D \subseteq \mathbf{R}^m$ be twice differentiable, and let $F'' : \mathbf{I}(D) \to \mathbf{I}^{m \times m}$ be an inclusion function for the Hessian matrix f''. Then $T_2 : \mathbf{I}(D) \to \mathbf{I}$ defined by
$$\begin{aligned} T_2(Y) = \ & f(c) + (Y-c)^T f'(c) \\ & + \tfrac{1}{2}(Y-c)^T F''(Y)(Y-c) \quad \text{for } Y \in \mathbf{I}(D), \end{aligned}$$

where $c = \text{mid } Y$ or any other point in Y, is called a *Taylor form function* (or shorter: *Taylor form*) for f of second order. Because of the Taylor formula of analysis, T_2 is an inclusion function for f. We say that F'' is bounded if a matrix $C \in \mathbf{I}^{m \times m}$ exists such that $F''(Y) \subseteq C$ for all $Y \in \mathbf{I}(D)$.

THEOREM 2 *If f is twice differentiable, and if f'' has a bounded inclusion function F'' then the Taylor form function, T_2, is of convergence order two.* □

If m is larger then it is better to avoid the explicit evaluation of $F''(Y)$ because of the many arithmetic operations one has to perform in order to obtain $T_2(Y)$. In such cases it might be better to compute $(Y-c)^T F''(Y)$ or $(Y-c)^T F''(Y)(Y-c)$ recursively as is suggested by McCormick (1983). For such a recursive computation the automatic differentiation arithmetic is appropriate as well, cf. Rall (1981). Even if $F''(Y)$ is needed explicitly it may be checked whether the recursive computation mentioned above is still faster. The tensor methods of Schnabel-Frank (1984) are also worth considering. Since all these techniques are not typical for the interval approach pursued by us we do not include a further discussion.

We again return to the meanvalue form (2.13)

$$T_1(Y, c) = f(c) + (Y-c)^T F'(Y),$$

but we now consider its dependence on the point c. We will describe a nice idea due to Baumann (1986) who provides formulas for $c \in Y$ such that $T_1(Y, c)$ gives optimum lower or upper bounds for $\Box f(Y)$ in a sense made precise below. The choice of $c = \text{mid } Y$ is still very popular because of the large saving in computational effort due to the symmetry of $Y - c$, cf. formulas (2.8), for example. Baumann's results which were derived in a more general setting are as follows: Let $Y = (Y_1, \ldots, Y_m)^T$, $Y_i = [x_i, y_i] \in \mathbf{I}$, $F'(Y) = (F_1'(Y), \ldots, F_m'(Y))^T$, $F_i'(Y) = [l_i, l_i']$. We define the "centers" $c^-, c^+ \in \mathbf{R}^m$ by

$$c_i^- = \begin{cases} y_i, & \text{if } l_i' \leq 0 \\ x_i, & \text{if } l_i \geq 0 \\ (l_i' x_i - l_i y_i)/(l_i' - l_i), & \text{otherwise,} \end{cases}$$

$$c_i^+ = \begin{cases} x_i, & \text{if } l_i'' \le 0, \\ y_i, & \text{if } l_i \ge 0, \\ (l_i x_i - l_i' y_i)/(l_i - l_i'), & \text{otherwise,} \end{cases}$$

for $i = 1, \ldots, m$. Then

$$\text{lb} T_1(Y, c) \le \text{lb} T_1(Y, c^-),$$
$$\text{ub} T_1(Y, c^+) \le \text{ub} T_1(Y, c)$$

for all $c \in Y$. These formulas mean that we may use c^- or c^+ instead of any other $c \in Y$ when we are interested in optimum lower or optimum upper bounds of f in Y, respectively, where $F'(Y)$ is assumed to be fixed. Our own numerical tests with global optimization problems show a reduction of the number of inclusion function evaluations by 10-20 percent when using c^- or c^+ instead of $c = \text{mid } Y$.

Another subject related to meanvalue forms which is still to be discussed is their generalizations for applications to *nonsmooth optimization*. A broad spectrum of mathematical programming problems can be rather easily reduced to the minimization of nondifferentiable functions without constraints or with simple constraints. The use of exact nonsmooth penalty functions in problems of nonlinear programming, maximum functions to estimate discrepancies in constraints, piecewise smooth approximation of technical-economic characteristics in practical problems of optimal planning and design, minimax compromise function in problems of multi-criterion optimization, all of these generate problems of nondifferentiable optimization. Thus, the objective function, f, of the optimization problem may look like

$$f(x) = \max\{f_1(x), \ldots, f_n(x)\},$$

where $f_i \in \mathbf{C}^1$, or, as a special case, like

$$f(x) = \| (f_1(x), \ldots, f_n(x)) \|$$

where the norm is the maximum norm. Objective functions arising from penalty methods are typically of the form

$$f(x) = \mu f_0(x) + \sum_{i=1}^{k} \max(0, g_i(x)) + \sum_{i=k+1}^{l} |h_i(x)|$$

where $f_0, g_i, h_i \in \mathbf{C}^1$ and $\mu > 0$ is a (reciprocal) penalty factor, cf. Sec. 1.4.

It turns out that the meanvalue form may also be used in such cases. One only has to replace the gradient by the generalized gradients which exist in these cases, cf. Rockafellar (1981). It is as easy or as difficult to find inclusions for the generalized gradient as for the gradient. Let $f : D \to \mathbf{R}$, $D \subseteq \mathbf{R}^m$ and $x \in D$. Furthermore let f be Lipschitz near x, that is, there exists an open neighborhood of x, say U_x, in which f satisfies a Lipschitz condition. It follows by a theorem of Rademacher that f is differentiable almost everywhere in U_x. Let Ω be the set of points in U_x at which f is not differentiable, and let S be any other set of Lebesgue measure 0. Then the *generalized gradient* of f at x is defined as

$$\partial f(x) = \mathrm{conv}\ \{\lim_{n \to \infty} f'(x_n) : x_n \to x, x_n \notin S \cup \Omega\}$$

where conv denotes the convex hull, cf. Clarke (1983). One can immediately see that $\partial f(x) = f'(x)$ when f is differentiable at x. Let $(x, y) \subseteq \mathbf{R}^m$ denote the open line segment between x and y. A theorem of Lebourg (1975) says that, if $y \in U_x$ with $(x, y) \subseteq U_x$ is given, then some $u \in (x, y)$ exists such that

$$f(y) - f(x) \in (y - x)^T \partial f(u). \qquad (2.14)$$

Locally, (2.14) can be approximated by means of the Lipschitz constant. Globally, (2.14) can be used to find inclusion functions of f of a meanvalue type explicitly: If $G(Y)$ is a - not necessarily bounded - box that contains $\partial f(u)$ for any $u \in Y$, then

$$F(Y) = f(c) + (Y - c)^T G(Y) \quad \text{for } Y \in \mathbf{I}(X), \qquad (2.15)$$

where $c = \mathrm{mid}\ Y$, is an inclusion function of f. Furthermore $G(Y)$ can be used for the *monotonicity test*: If only one component of $G(Y)$ does not contain zero, then f is strictly monotone with respect to the corresponding direction. I.e., if $G_i(Y)$ denotes the i-th component of $G(Y)$ then $0 < G_i(Y)$ or $G_i(Y) < 0$ implies that f is strictly monotonically increasing or decreasing, respectively, in Y with respect to x_i.

The optimality of Baumann's developing points c^- and c^+ as defined above is still preserved in case of (2.15).

Thus, practically, there seems to be no essential difference between the forms (2.13) and (2.15). There is one, however: (2.15) is not an inclusion function of second order if f is not differentiable. Still if f is differentiable in a neighborhood of the global minimizers then the algorithms we will be discussing in this monograph will finally process only boxes Y in which f is differentiable. Thus (2.15) is finally reduced to a meanvalue form of quadratic order if $G(Y)$ is chosen in an appropriate manner.

Example. Let $f(x) = x_1 + |x_2|$ for $x \in \mathbf{R}^2$. We get for the components $\partial f_i(x)$ of $\partial f(x)$,

$$\partial f_1(x) = 1, \quad \partial f_2(x) = \begin{cases} 1 & \text{if } x_2 > 0, \\ -1 & \text{if } x_2 < 0, \\ [-1, 1] & \text{if } x_2 = 0. \end{cases}$$

It is now not difficult to construct an inclusion function G of ∂f as follows,

$$G(Y) = \begin{cases} (1, 1)^T & \text{if } 0 < Y_2, \\ (1, -1)^T & \text{if } Y_2 < 0, \\ (1, [-1, 1])^T & \text{if } 0 \in Y_2, \end{cases}$$

for $Y \in \mathbf{I}(\mathbf{R}^2)$. Considering just $G(Y)$, one concludes that f is strictly monotonically increasing with respect to x_1, and strictly monotonically increasing or decreasing with respect to x_2 in Y if $0 < Y_2$ or $Y_2 < 0$, respectively. Finally,

$$F(Y) = f(c) + (Y - c)^T G(Y)$$

where $c \in Y$ is an inclusion function of f.

Summary. We sum up the contents of this section. There are two kinds of inclusion functions which can easily be constructed:

(1) Natural interval extensions,

(2) Centered forms:

(a) Meanvalue forms,

(b) Taylor forms (of second order).

Natural interval extensions may be used in general even if f is not differentiable.

Meanvalue forms may be used if f is differentiable, if f' has an inclusion function F' which is Lipschitz, and if $w(Y) \leq \frac{1}{2}$.

Meanvalue forms involving generalized gradients may be used if it cannot be decided from the outset whether f is differentiable or only generalized differentiable, and if $w(Y) \leq \frac{1}{2}$. Such an indeterminate situation occurs, for example, if $f(x) = \max(f_1(x), f_2(x))$ with $f_1, f_2 \in \mathbf{C}^1$. Then the differentiability properties of f at x cannot be determined before $f_1(x)$ and $f_2(x)$ are evaluated.

Taylor forms may only be used if a direct computation of the meanvalue form is not possible or if the Hessian inclusion $F''(Y)$ is already available and can be incorporated without difficulties. f has to be twice differentiable, f'' must have a bounded inclusion function F'' and $w(Y)$ should not be larger than $1/m$.

2.8 Improved Interval Hessian Matrices

The computation of $f''(x)$, that is, the Hessian matrix of f at x, is a frequently disputed topic. $f''(x)$ is simple to understand and to apply (from a theoretical point of view), but expensive to compute or to process when the dimension m is high and when $f''(x)$ is needed explicitly or when it is incorporated in terms such as $(x-c)^T f''(\xi)(x-c)$. A few references as to how to avoid the high costs were given in the last section. The use of interval tools does not change this situation where interval extensions $f(Y)$ of $f''(x)$ to a box Y occur. In this section we present a method whereby a substitute for $f''(Y)$ is constructed that will have as many non-interval entries as possible resulting in lower computational costs and sharper results. The idea is due to Hansen (1980), cf. also Rokne-Bao (1987).

Let again $f : D \to \mathbf{R}$, $D \subseteq \mathbf{R}^m$ and let f be in $\mathbf{C}^2$. Let $Y \in \mathbf{I}(D)$

Improved Interval Hessian Matrices

and $c \in Y$ be fixed. Expanding f about c the following is obtained,

$$f(x) = f(c) + (x-c)^T f'(c) + \frac{1}{2}(x-c)^T H(c, x, \Gamma)(x-c). \quad (2.16)$$

The matrix $H(c, x, \Gamma)$ has components $H_{ij}(c, x, \Gamma)$ which are defined in the following manner. First of all Γ is a lower triangular matrix consisting of the meanvalues γ_{ij} that arise during the Taylor expansion as sketched out in the sequel such that

$$\gamma_{ij} \in c_j \vee x_j \subseteq Y_j, \quad i = 1, 2, \ldots, m, \quad j = 1, 2, \ldots, i$$

and $\gamma_{ij} = 0$ otherwise. Then let

$$s_{ij}(x) = \frac{\partial^2 f(x)}{\partial x_i \partial x_j}, \quad i = 1, 2, \ldots, m, \quad j = 1, 2, \ldots, i.$$

The elements of the matrix $H(c, x, \Gamma)$ are now

$$H_{ij}(c, x, \Gamma) = \begin{cases} s_{ii}(x_1, \ldots, x_{i-1}, \gamma_{ii}, c_{i+1}, \ldots, c_m), & 1 \le i \le m, \ j = i \\ 2s_{ij}(x_1, \ldots, x_{i-1}, \gamma_{ij}, c_{i+1}, \ldots, c_m), & 1 \le i \le m, \ j < i \\ 0 & \text{otherwise,} \end{cases}$$

that is, a lower triangular matrix. The reason for this particular definition of the matrix $H(c, x, \Gamma)$ is related to the fact that the matrix should have as few interval variables as possible when evaluated in a sense to be defined below.

Since this expansion is not very familiar it is developed in greater detail for the case $m = 3$ around $c = (c_1, c_2, c_3)^T \in Y = (Y_1, Y_2, Y_3)^T$. The function f is first expanded as a function in x_3 around c_3 obtaining

$$\begin{aligned} f(x_1, x_2, x_3) &= f(x_1, x_2, c_3) + (x_3 - c_3) \frac{\partial}{\partial x_3} f(x_1, x_2, c_3) \\ &+ \frac{1}{2}(x_3 - c_3)^2 \frac{\partial^2}{\partial x_3^2} f(x_1, x_2, \gamma_{33}) \end{aligned} \quad (2.17)$$

where $\gamma_{33} \in c_3 \vee x_3$. Since $c_3, x_3 \in Y_3$ it also follows that $\gamma_{33} \in Y_3$.

The first two terms on the right side of (2.17) are now expanded around c_2 getting

$$\begin{aligned} f(x_1, x_2, c_3) &= f(x_1, c_2, c_3) + (x_2 - c_2) \frac{\partial}{\partial x_2} f(x_1, c_2, c_3) \\ &+ \frac{1}{2}(x_2 - c_2)^2 \frac{\partial^2}{\partial x_2^2} f(x_1, \gamma_{22}, c_3) \end{aligned} \quad (2.18)$$

and

$$\frac{\partial}{\partial x_3}f(x_1,x_2,c_3) = \frac{\partial}{\partial x_3}f(x_1,c_2,c_3) + (x_2 - c_2)\frac{\partial^2}{\partial x_3 \partial x_2}f(x_1,\gamma_{32},c_3) \quad (2.19)$$

where $\gamma_{22} \in c_2 \vee x_2 \subseteq Y_2$ and $\gamma_{32} \in c_2 \vee x_2 \subseteq Y_2$.

The process is repeated once more for the first two terms on the right side of (2.18) obtaining

$$\begin{aligned}f(x_1,c_2,c_3) &= f(c_1,c_2,c_3) + (x_1 - c_1)\frac{\partial}{\partial x_1}f(c_1,c_2,c_3) \\ &+ \frac{1}{2}(x_1 - c_1)^2 \frac{\partial^2}{\partial x_1^2}f(\gamma_{11},c_2,c_3)\end{aligned} \quad (2.20)$$

and

$$\frac{\partial}{\partial x_2}f(x_1,c_2,c_3) = \frac{\partial}{\partial x_2}f(c_1,c_2,c_3) + (x_1 - c_1)\frac{\partial^2}{\partial x_2 \partial x_1}f(\gamma_{21},c_2,c_3) \quad (2.21)$$

and for the first term on the right side of (2.19) getting

$$\frac{\partial}{\partial x_3}f(x_1,c_2,c_3) = \frac{\partial}{\partial x_3}f(c_1,c_2,c_3) + (x_1 - c_1)\frac{\partial^2}{\partial x_3 \partial x_1}f(\gamma_{31},c_2,c_3). \quad (2.22)$$

When equations (2.17) to (2.22) are combined the expansion

$$f(x) = f(c) + (x - c)^T f'(c) + \frac{1}{2}(x - c)^T H(c,x,\Gamma)(x - c) \quad (2.23)$$

is obtained. Here $\gamma_{ij} \in Y_j$ for $i = 1, 2, 3$ and $j \leq i$ and

$$H(c,x,\Gamma) = \begin{bmatrix} s_{11}(\gamma_{11},c_2,c_3) & 0 & 0 \\ 2s_{21}(\gamma_{21},c_2,c_3) & s_{22}(x_1,\gamma_{22},c_3) & 0 \\ 2s_{31}(\gamma_{31},c_2,c_3) & 2s_{32}(x_1,\gamma_{32},c_3) & s_{33}(x_1,x_2,\gamma_{33}) \end{bmatrix}.$$

Let now the matrix $H(c,Y)$ be defined by replacing the parameters x_j and γ_{ij} in $H(c,x,\Gamma)$ by the interval $Y_j (i = 1,2,3; j \leq i)$. From inclusion isotonicity it therefore follows that $H(c,x,\Gamma) \in H(c,Y)$. In some sense the matrix $H(c,Y)$ can be interpreted as a remodeled and atrophied interval Hessian matrix of f.

Improved Interval Hessian Matrices

From (2.23) it now follows that if $x \in Y$ then

$$f(x) \in \tilde{T}_2(Y) = f(c) + (Y-c)^T f'(c) + \frac{1}{2}(Y-c)^T H(c,Y)(Y-c) \quad (2.24)$$

and hence

$$\Box f(Y) \subseteq \tilde{T}_2(Y).$$

If the regular Taylor expansion was used to form the interval expansion which gave the Taylor form of the last section, $T_2(Y) = f(c) + (y-c)^T f'(c) + \frac{1}{2}(Y-c)^T f''(Y)(Y-c)$, then an interval Hessian matrix $f''(Y)$ would be used having more interval elements than $\tilde{T}_2(Y)$ resulting in poorer inclusions.

In order to find stationary points of f, interval Newton methods are applied to $g = f'$, cf. the next section. The interval Newton method requires inclusions for the first derivative of the gradient. This inclusion is best obtained using a similar expansion to (2.23): see Hansen (1968).

For the moment we drop the connection $g = f'$ and admit arbitrary vector-valued functions $g : D \to \mathbf{R}^m$ with $D \subseteq \mathbf{R}^m$. Since we again have to apply the meanvalue formula we first consider the component functions g_i of g instead of g itself.

Let $Y \in \mathbf{I}(D)$ and $x, c \in Y$. For $i = 1, \ldots, m$ we obtain:

$$g_i(x_1, \ldots, x_m) = g_i(x_1, \ldots, x_{m-1}, c_m)$$
$$+ \frac{\partial}{\partial x_m} g_i(x_1, \ldots, x_{m-1}, \xi_{im})(x_m - c_m)$$

where $\xi_{im} \in x_m \vee c_m$. Similarly, for $j = m-1, \ldots, 1$,

$$g_i(x_1, \ldots, x_{j-1}, x_j, c_{j+1}, \ldots, c_m) =$$
$$g_i(x_1, \ldots, x_{j-1}, \xi_{ij}, c_{j+1}, \ldots, c_m) + J_{ij}(c, x, \Xi_i)$$

where $\xi_{ij} \in x_j \vee c_j$ and

$$J_{ij}(c, x, \Xi) := \frac{\partial}{\partial x_j} g_i(x_1, \ldots, x_{j-1}, \xi_{ij}, c_{j+1}, \ldots, c_m)$$

with $\Xi_i := (\xi_{i1}, \ldots, \xi_{im})^T$. Substituting the first term at the right in each equation by the related formula recursively we get

$$g_i(x) = g_i(c) + (J_{i1}(c, x, \Xi_i) \ldots J_{im}(c, x, \Xi_i))(x - c). \quad (2.25)$$

We are now ready to consider the components $g_1, \ldots, g_m$ of g simultaneously. First we let the matrix $\Xi := (\Xi_1, \ldots, \Xi_m) = (\xi_{ij})$ collect all the meanvalues that have occurred. Then we define the matrix

$$J(c, x, \Xi) := (J_{ij}(c, x, \Xi_i)).$$

We note that $J(c, x, \Xi)$ is the Jacobian matrix of g at x if $c = x$ and $x_j = \xi_{ij}$ ($i, j = 1, \ldots, m$). The equations of (2.25) are written concisely as

$$g(x) = g(c) + J(c, x, \Xi)(x - c).$$

Let $J(c, Y)$ denote the natural interval extension of $J(c, x, \Xi)$ where x and the Ξ_i's are replaced by Y. Thus,

$$g(x) \in \tilde{T}_1(Y) := g(c) + J(c, Y)(Y - c) \quad \text{for all } x \in Y$$

which shows that $\tilde{T}_1(Y)$ is an inclusion function of g having components which are similar to the meanvalue form as defined in Sec. 2.7, but having a smaller number of intervals which results in better inclusions.

If $g = f'$ it is interesting to observe that $H(c, Y)$ is known when $J(c, Y)$ has been evaluated: In this case we have

$$g_i(x) = \frac{\partial f(x)}{\partial x_i},$$

$$\frac{\partial}{\partial x_j} g_i(x) = \frac{\partial^2}{\partial x_j \partial x_i} f(x) = \frac{\partial^2}{\partial x_i \partial x_j} f(x) = s_{ij}(x)$$

if $i \leq j$ which implies

$$\begin{aligned} J_{ij}(c, Y) &= \frac{\partial^2}{\partial x_i \partial x_j} f(Y_1, \ldots, Y_j, c_{j+1}, \ldots, c_m) \\ &= s_{ij}(Y_1, \ldots, Y_j, c_{j+1}, \ldots, c_m). \end{aligned}$$

The connection between $H(c, Y)$ and $J(c, Y)$ is thus given by

$$H_{ij}(c, Y) = \begin{cases} J_{ij}(c, Y) & \text{if } i = j \\ 2 J_{ij}(c, Y) & \text{if } i < j. \end{cases}$$

2.9 Interval Newton Methods

Interval Newton methods are excellent methods for determining *all zeros* of a continuously differentiable vector-valued function $\phi : X \to \mathbf{R}^m$ where $X \in \mathbf{I}^m$. These methods are important tools for nonlinear optimization problems since they can be used for computing all critical points of ϕ by applying the methods to $J_\phi(x)$, or for solving the Kuhn-Tucker or John conditions in constrained optimization.

The interval Newton method was introduced by Moore (1966), refinements and discussions are due to Krawczyk (1969), Nickel (1971), Hansen (1978b), Hansen-Sengupta (1981), Hansen-Greenberg (1983), Alefeld-Herzberger (1983), Krawczyk (1986), Neumaier (1985), Wolfe (1980), Schrempp (1984), Bao-Rokne (1987) and many others.

Let $x, y \in X$ and $\phi = (\phi_1, \ldots, \phi_m)^T$ be expanded componentwise by the meanvalue formula at x,

$$\phi(y) = \phi(x) + J(\sigma)(y - x),$$

where

$$J(\sigma) = (\phi'_1(\sigma_1), \ldots, \phi'_m(\sigma_m))^T$$

for a matrix $\sigma = (\sigma_1, \ldots, \sigma_m), \sigma_i \in \mathbf{R}^m$ and $\sigma_i \in x \lor y$. We define

$$J(Y), \quad Y \in \mathbf{I}(X)$$

a bit outside our usual convention as the natural interval extension of J to Y^m, that is, each σ_i is replaced by $Y, i = 1, \ldots, m$. Owing to the definition of $J(\sigma)$ we obtain

$$J(Y) = J_\phi(Y),$$

such that $J(Y)$ is nothing but a natural interval extension of the Jacobian matrix $J_\phi(x)$ to Y. Note that $J(\sigma)$ is *not* a Jacobian matrix. If $y = \xi$ is any zero of ϕ in X then

$$J(\sigma)(x - \xi) = \phi(x).$$

This equation leads to interval Newton methods, in the the same manner as we get non-interval Newton methods in the non-interval case, cf. the iteration statement given below.

Before the method is specified we have to define what is meant by solving a system of linear interval equations. An unfortunate notation is widely used to describe this situation since it uses the notation of interval arithmetic in a doubtful manner. This can lead to misunderstandings. I.e., let $A \in \mathbf{I}^{m \times m}$, $B \in \mathbf{I}^m$ then the *solution of the linear interval equation (with respect to x or X)*

$$Ax = B \text{ or } AX = B$$

is not an interval vector X_0 that satisfies the equation, $AX_0 = B$, as one would expect. The solution is defined as the set

$$X = \{x \in \mathbf{R}^m : ax = b \text{ for some } a \in A, b \in B\}.$$

Thus, for example, the solution of the linear interval equation

$$[1,\ 2]x = [1,\ 2]$$

is $X = [1/2,\ 2]$. If we multiply, for comparison, $[1,\ 2]$ and X we get

$$[1,\ 2]X = [1,\ 2][1/2,\ 2] = [1/2,\ 4].$$

In general, the solution set is not a box. It is therefore the aim of interval arithmetic solution methods to find at least a box which contains the solution set. Accordingly, if $c \in \mathbf{R}^m$, then the solution of the linear interval equation

$$A(x - c) = B \text{ or } A(X - c) = B$$

with respect to x or X is defined to be the set

$$X := c + Y := \{c + y : y \in Y\}$$

where Y is the solution of the interval equation $Ay = B$.

The following prototype algorithm aims to determine the zeros of $\phi : X \to \mathbf{R}^m$ in $X \in \mathbf{I}^m$.

The Interval Newton Algorithm

1. Set $X_0 := X$.

2. For $n = 0, 1, 2, \ldots$

   (i) choose $x_n \in X_n$,

   (ii) determine a superset Z_{n+1} of the solution Y_{n+1} of the linear interval equation with respect to Y

   $$J(X_n)(x_n - Y) = \phi(x_n),$$

   (iii) set $X_{n+1} := Z_{n+1} \cap X_n$.

Since we use it later we emphasize that *one iteration of the interval Newton Algorithm* is just the execution of (i), (ii) and (iii) for a particular value of n.

The interval Newton methods are distinguished by how (ii) is solved for Y_{n+1}. Convergence properties exist under certain assumptions. The following general properties are useful for understanding the principle of application of the algorithm:

1. If a zero, ξ, of ϕ exists in X then $\xi \in X_n$ for all n, cf. Moore (1966).

   This means that no zero is ever lost! This implies:

2. If X_n is empty for some n then ϕ has no zeros in X, cf. Moore (1966).

3. If $Y_{n+1} \subseteq X_n$ for some n then a zero of ϕ exists in X, cf. Bao-Rokne (1987).

The next section discusses a very effective practical realization of the Newton algorithm.

2.10 The Hansen-Greenberg Realization

To date, the most promising realization of the interval Newton method is that developed by Hansen-Greenberg (1983). Many numerical tests have demonstrated its effectiveness. Neumaier's (1985) realization is also remarkable.

Considering the interval Newton Algorithm one notices that its quality depends mainly on the right choice of a method for solving the equation

$$J(X_n)(x_n - Y) = \phi(x_n) \qquad (2.26)$$

in the sense defined in the last section. The method of Hansen-Greenberg uses a combination of iterative and algebraic steps for solving such linear equations. The goal of their method is to delay the unavoidable matrix inversion or related costly operations as long as possible as is also the case in quasi Newton methods. The method is based upon

(A) a preconditioning step,

(B) a relaxation procedure,

(C) a local iteration procedure,

(D) a Gaussian elimination procedure.

Since we discuss just one iteration of the Hansen-Greenberg variant in the sequel we suppress the indices n when writing down the formulas that occur in the n-th iteration. That is, we write

$$J(X)(x - Y) = \phi(x) \qquad (2.27)$$

instead of

$$J(X_n)(x_n - Y) = \phi(x_n)$$

and, accordingly, we search for a superset Z of the solution set of (2.27), where $X, J(X), x$ and $\phi(x)$ are given. The solution set of (2.27) is also denoted by Y. If we want to refer to the former, original box X, we are more likely to avoid misunderstandings if we speak of the initial box X_0, reminding us that the box X_0 was defined as the original box X. We start with $x = $ mid X. During the iteration, mainly by the procedures (B) and (D), only x and X will be updated, but not $J(X)$. By "updating" it is meant that the result of a formula or dependent variable is improved without a complete recomputation of that formula, etc. The individual steps of the Hansen-Greenberg variant are now discussed in some detail.

(A) The preconditioning step

It was argued in Hansen-Smith (1967) that (2.27) was best solved by pre-multiplying by an approximate inverse of mid $J(X)$. If the approximate inverse is B then we obtain

$$BJ(X)(x - Y) = B\phi(x)$$

or

$$M(x - Y) = b \qquad (2.28)$$

where $M = BJ(X)$ and $b = B\phi(x)$. In this manner the system has been modified to a system that is almost diagonally dominant provided the widths of the Jacobian entries are not too large. This was also discussed by Miller (1972a). He showed that if $Y = \{z \mid m(x - z) = b, m \in M\}$ then the superset Z of Y obtained by solving (2.28) by Gaussian elimination without pivoting satisfies $w(Z) = w(Y) + w(J(X))^2$.

Such systems are also amenable to Gauss-Seidel type iterations because of the likely diagonal dominance. This will be discussed below.

It is obvious that the solution set of (2.28) contains the solution set of (2.27) such that no solution is lost in the above transformation. The only thing we have to do now is to solve the linear interval equation (2.28).

(B) The relaxation procedure

Because of the properties of the interval Newton algorithm cited in the last section we know that all zeros of ϕ lie in X (where this X abbreviates the former X_n), and therefore in the solution set of (2.28). The relaxation procedure tries to update X and x in order to make the elimination procedure, which will finally be applied, more effective. By updating it is meant here that a smaller X is obtained as well as an x nearer to a zero of ϕ such that x is a better developing point for the methods (C) and (D) which follow later. It can also happen that when X is made smaller it is split into two or more disconnected boxes, containing all solutions such that basically the further procedures need only be applied to these subboxes. In order to avoid an exponential increase of the number of subboxes the further steps are applied to the hull of the subboxes in such cases. A splitting into two subboxes is only done when the current iteration of the Newton method is finished.

The relaxation procedure for linear interval equations was developed in Hansen-Sengupta (1981). It consists of the application of the Gauss-Seidel iteration procedure (see for example Conte-de Boor (1980)) in an interval context (see also the discussion of related methods in Alefeld-Herzberger (1983)). *This relaxation procedure is here used to solve the preconditioned set of equations* (2.28) and it is expected that the procedure will be efficient since the coefficient matrix M will in most cases contain the identity matrix I due to its construction as $BJ(X)$, although this is not guaranteed since we only required that B should be an approximate inverse of mid $J(X)$. It should also be noted that the matrix M is kept constant throughout the relaxation steps (i.e. $M = BJ(X)$) whereas the vector X is updated.

In the relaxation procedure the equation $M(x - Y) = b$ is solved for the i-th component Y_i obtaining a superset of Y_i,

$$Z_i = x_i + M_{ii}^{-1}(\sum_{\substack{j=1 \\ j \neq i}}^{m} M_{ij}(x_j - X_j) - b_i) \qquad (2.29)$$

where x_j, etc., denotes the j-th component of x, etc. This interval is immediately used to intersect and update the i-th component X_i,

$$X_i := X_i \cap Z_i. \qquad (2.30)$$

This calculation is performed for all $i, 1 \leq i \leq m$, first for the indices where $0 \notin M_{ii}$ and then for the remaining indices where $0 \in M_{ii}$. This strategy results from the observation that the updating (2.30) with components X_i where $0 \notin M_{ii}$ improves (makes smaller) all the components Z_j via formula (2.29). This does however not hold for components X_i with $0 \in M_{ii}$.

If the intersection (2.30) is empty for some i then it follows from the properties cited in the last section that there is no solution in

X and thus no solution in the initial box X_0. Therefore the whole interval Newton algorithm is to be terminated reporting

- no zero of ϕ in X_0.

When the intersection is not empty then the iteration process continues with the next component and the updated X_i's.

If $0 \in M_{ii}$ and if

$$0 \in \sum_{\substack{j=1 \\ j \neq i}}^{m} M_{ij}(x_j - X_j) - b_i$$

then we set $Z_i = (-\infty, \infty)$. In this case the intersection (2.30) does not result in a narrowing of X_i; hence no useful information is obtained.

If

$$0 \notin \sum_{\substack{j=1 \\ j \neq i}}^{m} M_{ij}(x_j - X_j) - b_i$$

and $0 \in M_{ii}$ then Z_i consists of two non-overlapping semi-infinite intervals separated by an open set (gap) according to the extended interval arithmetic division given in (2.7). The intersection $Z_i \cap X_i$ may now be empty or consist of one or two intervals. In the first two cases the iteration proceeds as for the case $0 \notin M_{ii}$.

If the intersection results in two intervals then the box may be split normal to this coordinate direction. It might be impractical when the box is split with respect to several coordinate directions, thus resulting in a proliferation of subboxes as mentioned before. A splitting is therefore only done once during the iteration, i.e. vertical to the direction of the largest gap at the end of the iteration. In practice the algorithm therefore has to keep track of both the gap and the index of the coordinate where it occurs.

The gaps are also not used right away; they are saved until the other techniques have been employed to narrow the current box. It also turns out that the extended interval arithmetic calculations provide a wider gap when x is a poor approximation for a zero of ϕ. For this reason a gap is calculated prior to using the remaining techniques for shrinking the width of a box.

Combining the above steps the following intermediate procedure is obtained.

Relaxation procedure algorithm

1. For $i = 1, 2, \ldots, m$

    if $0 \notin M_{ii}$ then

    (a) set
    $$Z_i := x_i + M_{ii}^{-1}\left(\sum_{\substack{j=1 \\ j \neq i}}^{m} M_{ij}(x_j - X_j) - b_i \right)$$

    and
    $$X_i := X_i \cap Z_i.$$

    (b) If $X_i = \emptyset$ then terminate and report
    - no solution in X.

2. For $i = 1, 2, \ldots, m$

    if $0 \in M_{ii}$ then

    (a) set
    $$Z_i := x_i + M_{ii}^{-1}\left(\sum_{\substack{j=1 \\ j \neq i}}^{m} M_{ij}(x_j - X_j) - b_i \right)$$

    and
    $$X_i := X_i \cap Z_i.$$

    (b) If $X_i = \emptyset$ then terminate and report
    - no solution in X.

    (c) If X_i has a gap then save i and gap.

(C) The local iteration procedure

The relaxation method is a rather robust step which is applicable also in unfavorable cases. Thus it is a slow step. In order to apply methods which are more accurate and faster, like the elimination procedure (D), it is necessary to be as near as possible to a solution. For this reason a shift of $x = \text{mid } X$ in the direction of a solution, where X is the resulting interval of the procedure (B), is desirable. This shift is done by the local iteration procedure discussed here. If the shift is successful, i.e. $\| \phi(x) \| < 10^{-3}$, then the elimination procedure (D) is applied. Otherwise, the relaxation procedure is applied in a simplified manner, that is, only part 1 of the algorithm is used.

For the shift we simply use a non-interval quasi Newton method,

$$\begin{aligned} x_0 &:= \text{mid}X, \\ x_{n+1} &:= x_n - B\phi(x_n) \end{aligned}$$

for $n = 0, 1, \ldots, s$. (Here the x_n denote again the iterates of x and not the coordinates of the vector x.) The number s is determined implicitly when the termination criterion is satisfied, that is,

$$\| \phi(x_{s+1}) \| < 10^{-3} \text{ or } \| \phi(x_{s+1}) \| > \tfrac{1}{2} \| \phi(x_s) \|.$$

The norm used is the Euclidean norm. We set the updated value of x to whichever x_s and x_{s+1} yields the smaller norm.

If a step from x_n to x_{n+1} leads out of the current box X, a point on the boundary of X which is on the line connecting x_n and x_{n+1} is used as a replacement for x_{n+1}.

(D) The elimination procedure

The elimination method (due to Gauss) applied to a uniquely solvable system of linear equations gives the exact solution in a finite number of arithmetic operations assuming exact arithmetic. The elimination method is thus an ideal means of solving equations. The method may, however, degrade if rounding errors are involved or if interval arithmetic is used. In both cases the method works excellently under certain conditions. Since these conditions are rather restricted or troublesome to verify computationally, especially in the interval case, the elimination procedure will only be applied if $\| \phi(x) \|$ is sufficiently small. This condition indicates that x is near a zero and that, considering for a moment the equation (2.28) to be solved with respect to Y,

$$BJ(X)(x - Y) = B\phi(x),$$

it is plausible that $M = BJ(X)$ is near the identity matrix such that the elimination procedure may not require the search for pivots, thus saving row and column permutations.

If the elimination procedure is successful and if X is made smaller then the conditions for applying the elimination procedure are improved and the procedure is applied again. Otherwise we return to the relaxation procedure or terminate the current iteration of the Newton procedure.

The elimination procedure uses a LU decomposition of M without row or column permutations. This means, first in case of non-interval matrices $A \in \mathbf{R}^{m \times m}$, that a lower triangular matrix L and an upper triangular matrix U are determined such that

$$A = LU.$$

The necessary steps can be found in any textbook on numerical analysis cf. Conte-Boor (1980), p. 161. A LU decomposition of the interval matrix M means that those arithmetical operations are applied to M which would be used if M were a non-interval matrix. Owing to interval arithmetic properties the equality $M = LU$ cannot be obtained, only the inclusion

$$M \subseteq LU.$$

Thus, $LU(x-Y) = b$ instead of $M(x-Y) = b$ is solved. No zero of ϕ is lost by this weakening, however. As already mentioned we terminate the LU decomposition when a division through an interval containing 0 occurs. The use of the extended division would be possible, but this would worsen the conditions. In these cases we return, as mentioned, to the relaxation procedure.

The details of the procedure are given by the following steps, cf. Hansen-Sengupta (1983):

Elimination procedure

1. If $0 \in M_{ii}$ for some i, terminate.

2. Try to obtain the LU decomposition of M, that is, $M \subseteq LU$. Terminate when not successful.

3. Solve $Lw = b$ with respect to w. Let $W \in \mathbf{I}^m$ be a box that includes the solution set. Terminate if a division through a box containing 0 occurs.

4. Solve $U(x - Y) = W$ with respect to Y as follows:

    (a) Set $i := m$.

    (b) Terminate if $0 \in U_{ii}$. Compute $Z_i := x_i - (W_i - \sum_{j=i+1}^{m} U_{ij}(x_j - X_j))/U_{ii}$ (Z_i, x_i, etc. denote here the coordinates of Z, x, etc.). Z is a box containing the solution set.

    (c) Set $X_i := X_i \cap Z_i$ (updating of X).

    (d) Set $i := i - 1$.

    (e) If $i \neq 0$ go to (b).

5. End.

The order of application of the procedures (A) - (D)

The manner in which the steps and procedures (A) - (D) are combined can be seen from Alg. 1 below. It is, nevertheless, appropriate to describe some of the reasons for the particular choice of order of sub-algorithms in a less formal language.

The complete procedure for solving $J(X)(x - Y) = b$ with respect to Y within one iteration of the Hansen-Greenberg realization consists of the following items:

1. The preconditioning step transforms this equation to $M(x - Y) = b$.

2. The relaxation procedure is applied in order to improve X and x which are updated by $X := X \cap Z$ where Z is a superset of the solution set of $M(x - Y) = B$ and by $x := \text{mid } X$. Gaps may also be found and saved in order that they may be used for the final splitting in item 7.

3. The local iteration procedure is supposed to improve x (that is, finding an x nearer to a solution of $\phi(x) = 0$) and thus $b = B\phi(x)$.

4. If $\| \phi(x) \| < 10^{-3}$ the elimination procedure is applied for improving X. Three possibilities can occur:

    (i) The improvement of X is significant. By this we mean that the ratio of old box width to new box width is smaller than 0.9. (This threshold results from computational experience and should be higher the larger m is and the more expensive the recalculation of $J(X)$ is.) In this case the elimination procedure is applied again (that is, go back to item 4).

    (ii) The improvement of X is not significant. In this case the further improvements are skipped and the splitting of X - see item 6 - is done.

    (iii) The elimination operations require a division by an interval containing zero. The elimination procedure is then stopped and the simplified relaxation procedure - see item 5 - is applied.

    If $\| \phi(x) \| \geq 10^{-3}$ the simplified relaxation procedure is applied - see item 5.

5. The *simplified relaxation* procedure performs (2.29), that is the computation of the Z_i only if $0 \notin M_{ii}$. Therefore only part 1 is executed in the related algorithm for the relaxation procedure. The Z_i with $0 \in M_{ii}$ are not calculated again since they were already used to find gaps. If X improves significantly (cf. 4(i)), we set $x := \text{mid } X$ then compute $b := B\phi(x)$ and the simplified relaxation procedure is repeated (that is, go back to 5.).

If X does not improve significantly then X is split as in item 6 below.

6. The splitting of the box is executed at the end when further improvement by the other steps is unlikely. There are 2 possibilities:

    (i) If gaps were found in item 1 and not eliminated by the further improvements of X then a largest gap is used to split the box. More precisely if this largest gap occurs in direction i the box is split in the i-th direction. (If the boxes have edges which are very different in size, one can use the relatively largest gap which results from comparing the ratio gap width to edge width in each direction.)

    (ii) If no gaps were found in item 1 then the box is split at the midpoint of a largest component.

    The bisected boxes are put on a stack where they wait for further processing. The box with the largest width is taken from the stack and is denoted by X and a new iteration is done with this X after checking 7. If the stack has no boxes that can be chosen as a new X then ϕ has no solution at all in the initially given area X_0 and the algorithm is terminated.

7. Termination criteria. If not terminated then start a new iteration (i.e. go to 1.).

In later sections we will incorporate *one iteration* of the interval Newton method into optimization algorithms. By one iteration we then mean an iteration as described above omitting the artificial bisection performed in item (ii) of item 6 since this bisection will be executed by the optimization algorithm.

The reasons for using exactly the above sequence of algorithmic steps are primarily given on the basis of extensive numerical experience. One can make this choice of ordering plausible, as we did, but a serious proof that this order is the best order is missing.

We now give the detailed steps of the interval Newton method where the input parameters are the function ϕ, an inclusion function

Φ of ϕ, the starting box X in which the solutions (zeros of ϕ) are to be found and some termination parameters.

Interval Newton Algorithm (after Hansen-Greenberg):

ALGORITHM 1

1. Let X be given.

2. Initialize list $L = (X)$.

3. Calculate $\Phi(X)$. If $0 \notin \Phi(X)$ then go to 21.

4. Calculate $J(X)$.

5. Calculate B, an approximate inverse of mid $J(X)$.

6. Set x equal to the midpoint of X, that is $x = \text{mid } X$.

7. Calculate $M := BJ(X)$, and $b := B\phi(x)$.

8. Perform the relaxation procedure algorithm. If some $X_i = \emptyset$ then go to 21.

9. Perform the local iteration procedure to improve x.

10. If $\| \phi(x) \|$ is not sufficiently small (for applying the elimination procedure) then go to 16.

11. Perform the LU decomposition of M if possible. If LU decomposition is not possible go to 16.

12. Calculate $b := B\phi(x)$.

13. Perform the remaining steps of the elimination procedure (the LU decomposition has already been terminated successfully). If a division by an interval containing 0 occurs go to 16.

14. If $X = \emptyset$ then go to 21.

15. If X improved significantly in Step 13 set $x := \text{mid } X$ and go to 12. Otherwise go to 18.

16. Perform the simplified relaxation procedure. If $X = \emptyset$ then go to 21.

17. If X improved significantly in Step 16 set $x := \text{mid } X$, compute $b := B\phi(x)$ and return to 16.

18. If the gaps together with their coordinate directions were saved in Step 8 then

    - update the gaps (they could increase or even vanish by the continued shrinking process $X_i := X_i \cap Z_i$). Use the coordinate direction with the largest gap for splitting X using the gap obtaining boxes V_1 and V_2. This gap is no longer part of V_1 or V_2. (The remaining gaps are still included.) Go to 20.

19. If X did not improve significantly in Steps 8 or 13 or 16 then

    - choose a coordinate direction v parallel to which $Y_1 \times \ldots \times Y_m$ has an edge of maximum length, i.e. $v \in \{i : w(Y) = w(Y_i)\}$. Bisect Y vertical to direction v getting boxes V_1, V_2 such that $Y = V_1 \cup V_2$.

20. Enter V_1 and V_2 onto the list.

21. Remove X from the list. If list is empty, terminate and report

    - no solutions.

22. Apply termination criteria. For example: (i) If $\| \Phi(X) \| < \epsilon$ for all boxes X of the list then terminate. Or: (ii) If $w(X) < \tilde{\epsilon}$ for all boxes X of the list then terminate since a continuation seems to be unsuccessful. Etc.

23. Set X to be the box of the list with the largest width.

24. Return to 3.

2.11 Numerical Examples Using the Interval Newton Method

The interval Newton method described in Alg. 1 was implemented in Fortran and some test cases were run. The description of the results uses the same notation as in Hansen-Greenberg (1983) where

n_1 = the number of iterations of the interval Newton algorithm,
= the number of $\Phi(X)$ evaluations,
≥ the number of $J(X)$ evaluations,
≥ the number of inverse matrix evaluations,
≥ the number of interval arithmetic matrix-matrix products,
n_2 = the number of $\phi(x)$ evaluations,
= the number of interval arithmetic matrix-vector products,
= $n_3 + n_5 + n_6$,
n_3 = the number of local iterations,
n_4 = the number of interval arithmetic LU decomposition attempts (failing and succeeding),
n_5 = the number of executions of the elimination procedure,
n_6 = the number of executions of the relaxation procedure for those i with $0 \notin M_{ii}$,
n_7 = the number of executions of the relaxation procedure for those i with $0 \in M_{ii}$,
s = the significant improvement factor. This means that a box is considered to be improved significantly by the relaxation or elimination procedure if the ratio of new box width to old box width is smaller than s,
$E \pm n$ = $10^{\pm n}$.

Alg. 1 was then tested on three examples.

Example 1. In Hansen-Greenberg (1983) the Broyden banded function was used to test Alg. 1. This function is defined by

$$\phi_i(x) = x_i(2 + 5x_i^2) + 1 - \sum_{j \in J_i} x_j(1 + x_j) \quad (i = 1, \ldots, m),$$

where $J_i = \{j : j \neq i, \max(1, i-5) \leq j \leq \min(m, i+1)\}$. The function

Numerical Examples Using the Interval Newton Method

was chosen by Hansen-Greenberg (1983) for easy programmability in an arbitrary dimension.

Alg. 1 was executed for this function with $m = 3$ and with four different starting boxes $X = ([a,\ b], [a,\ b], [a,\ b])^T$ where $[a,\ b]$ are given in the table below.

The significant width improvement factor was $s = 0.9$ and for termination it was required that each box on the list had width less than 10^{-8}.

The statistics for the solution process for $\phi(x) = 0$ with the different choices of $[a,\ b]$ were

	Test 1	Test 2	Test 3	Test 4
$[a, b]$	$[-1, 1]$	$[-20, 15]$	$[-3, 2]$	$[-1, 2]$
n_1	15	33	23	16
n_2	64	169	212	53
n_3	19	32	29	21
n_4	1	0	0	2
n_5	17	0	0	4
n_6	28	137	183	28
n_7	1	0	0	0

There were 2 adjacent boxes on the final list in Test 1 and the union of these boxes was the box

$$\begin{pmatrix} [-.42830257, & -.42830256] \\ [-.47656629, & -.47656628] \\ [-.47656629, & -.47656628] \end{pmatrix}.$$

For the final boxes X on the final list we also have

$$\| \phi(X) \| \leq 2 \cdot 10^{-7}$$

where $\phi(X)$ is the natural interval extension of ϕ to X and

$$\| \phi(X) \| = \max_{y \in \phi(X)} \| y \|.$$

Test 2 also resulted in two final boxes, whereas Tests 3 and 4 had only one final box. In all the tests the results were similar to the results obtained in Test 1.

This example also showed the relative frequencies of execution of the different components of the algorithm.

Example 2. In Moré-Cosnard (1979) the nonlinear integral equation

$$u(t) + \int_0^1 H(s,t)(u(s) + s + 1)^3 ds = 0,$$

with

$$H(s,t) = \begin{cases} s(1-t), & s \leq t, \\ t(1-s), & s \geq t, \end{cases}$$

was discretized by considering the equation at the points $t = t_k$, $k = 1, \ldots, m$, and then replacing the integral by an m-point rectangular rule based on the points $\{t_k\}$. The resulting system of equations in the unknowns $x_k = u(t_k)$ was defined by

$$\phi_k(x) = x_k + \frac{1}{2}\{(1 - t_k)\sum_{j=1}^{k} t_j(x_j + t_j + 1)^3$$
$$+ t_k \sum_{j=k+1}^{m} (1 - t_j)(x_j + t_j + 1)^2\} \quad (2.31)$$

where $x_0 = x_{m+1} = 0$, $t_j = jh$, and $h = (m+1)^{-1}$.

This is the standard finite difference approach to solving nonlinear integral equations where the solutions of system $\phi(x) = 0$ provide approximations to the values of $u(t)$ at $t_0, \ldots, t_m$.

The final boxes on the list include the solutions to the finite difference equations (2.31) but not necessarily to the original integral equations. However, using known analytic techniques it is also possible to estimate the error committed by the finite difference approach and thus provide inclusions to the values of $u(t)$ at $t_0, \ldots, t_m$ as well.

Alg. 1 was executed for this function for a varying m with an initial box having components $X_i = [-4, 5]$, $i = 1, \ldots, m$, in each case. The significant width improvement factor was $s = 0.0$ and for termination it was required that each box on the list had width less than 10^{-6}.

Numerical Examples Using the Interval Newton Method

The statistics resulting from solving this problem for $m = 4, 5, 6, 7$ were

	$m = 4$	$m = 5$	$m = 6$	$m = 7$
n_1	1	1	1	1
n_2	20	22	25	27
n_3	1	1	1	1
n_4	0	0	0	0
n_5	0	0	0	0
n_6	19	21	24	26
n_7	0	0	0	0

In each case only one final solution box X remained. These were for $m = 4$:

$$\begin{pmatrix} [-.876487E - 01, & -.876483E - 01] \\ [-.147771E + 00, & -.147770E + 00] \\ [-.168620E + 00, & -.168619E + 00] \\ [-.130817E + 00, & -.130816E + 00] \end{pmatrix}$$

with $\| \phi(X) \| < 0.86 \times 10^{-6}$,

for $m = 5$:

$$\begin{pmatrix} [-.750223E - 01, & -.750216E - 01] \\ [-.131976E + 00, & -.131975E + 00] \\ [-.164849E + 00, & -.164848E + 00] \\ [-.164665E + 00, & -.164664E + 00] \\ [-.117418E + 00, & -.117417E + 00] \end{pmatrix}$$

with $\| \phi(X) \| < 0.2 \times 10^{-5}$,

for $m = 6$:

$$\begin{pmatrix} [-.654636E - 01, & -.654632E - 01] \\ [-.118166E + 00, & -.118165E + 00] \\ [-.154627E + 00, & -.154626E + 00] \\ [-.169992E + 00, & -.169991E + 00] \\ [-.157270E + 00, & -.157269E + 00] \\ [-.106031E + 00, & -.106030E + 00] \end{pmatrix}$$

with $\|\phi(X)\| < 0.14 \times 10^{-5}$,
for $m = 7$:

$$\begin{pmatrix} [-.580146E-01, & -.580143E-01] \\ [-.106539E+00, & -.106538E+00] \\ [-.143383E+00, & -.143382E+00] \\ [-.165632E+00, & -.165631E+00] \\ [-.169319E+00, & -.169318E+00] \\ [-.148908E+00, & -.148907E+00] \\ [-.964312E-01, & -.964310E-01] \end{pmatrix}$$

with $\|\phi(X)\| \leq 0.12 \times 10^{-5}$.

We note that in this example the interval Newton method only required one iteration for each of the cases $m = 4, 5, 6, 7$ and that the number of inner iterations only increased slowly with m.

Example 3. A very simple, but troublesome, global optimization problem was devised to test the effectiveness of this implementation of the interval Newton method for finding a continuum of zeros.

The problem was simply: Find all the global minimizers of $f(x) = \frac{1}{2}x_1^2 x_2^2$ in a given box X.

It is immediately obvious that the set of global minimizers is formed from $x_1 = 0$ and $x_2 = c$ together with $x_1 = d$ and $x_2 = 0$ where c, d are arbitrary, but restricted by the box X.

The interval Newton method was applied to the gradient of f in order to solve this problem thus obtaining a list of final boxes containing all the zeros of the gradient.

The gradient of f is

$$\frac{\partial f(x)}{\partial x_1} = x_1 x_2^2 = \phi_1(x)$$

$$\frac{\partial f(x)}{\partial x_2} = x_1^2 x_2 = \phi_2(x).$$

Alg. 1 was therefore applied to solve $\phi(x) = 0$ in the box $X = ([-1.0, \ 0.5], [-0.4, \ 0.5])^T$ with the significant width improvement factor $s = 0.9$. It was also required that the boxes on the final list had width less than 10^{-1} (knowing *a priori* that the number of boxes would then be of the order of 25-30).

The results were

n_1	94
n_2	107
n_3	41
n_4	37
n_5	0
n_6	66
n_7	94

There were 25 boxes on the final list. Two representative boxes were

$$X_2 = \begin{pmatrix} [-.91E+00, & -.81E+00] \\ [-.47E-01, & +.50E-01] \end{pmatrix}$$

with $\|f(X_2)\| \leq .10E - 2$ and

$$X_{16} = \begin{pmatrix} [-.63E-01, & +.31E-01] \\ [-.63E-02, & +.50E-01] \end{pmatrix}$$

with $\|f(X_{16})\| \leq .49E - 5$.

Chapter 3

Global Unconstrained Optimization

3.1 Introduction

In this chapter we consider the *global unconstrained optimization* (more precise: *minimization*) *problem*: Let $\mathbf{R}$ be the set of reals, let $X \subseteq \mathbf{R}^m$ be a compact right parallelepiped parallel to the axes (abbreviated as a *box* in the sequel), $f : X \to \mathbf{R}$ any function and $\Box f(X)$ the range of f over X, that is the set of all function values, $\Box f(X) = \{f(x) : x \in X\}$. The *global "minimum"* inf $(\Box f(X))$ is denoted by f^* if it exists in $\mathbf{R}$, the set of *global minimizers* (or *global minimum points*) by X^*. Since we do not assume the continuity of f at first, we use "inf" instead of "min". The problem mentioned is usually described by

$$\text{minimize } f(x) \quad \text{subject to } x \in X. \tag{3.1}$$

We will describe methods which are able to determine

(a) f^*, or

(b) f^* and X^*.

The numerical realization of these methods will be such that - depending on the properties of f -

- lower and upper bounds of f^*,

- arbitrarily good lower and upper bounds of f^*,

- inclusions of X^*,

- arbitrarily good inclusions of X^*

will be produced.

If the methods are applied to $-f$ then the methods will obviously determine the *global maximum* and the set of *global maximizers* of f over X.

We will present 3 algorithms for solving (3.1). The first algorithm, due to Moore (1966) and Skelboe (1974), only aims to determine f^*. The second algorithm, due to Ichida-Fujii (1979), is very similar to the Moore-Skelboe algorithm and it aims to determine f^* and X^*. This algorithm has the problem that convergence to X^* is not generally ensured. Since it is still frequently used we include it into our considerations and we discuss its properties. Finally, the algorithm of Hansen (1979), (1980) has just f^* and X^* as the solution set and it is therefore superior to the algorithm of Ichida-Fujii.

The 3 algorithms are based upon the *branch and bound principle*. By this we mean that:

(a) The whole area X is not searched uniformly for the global minimizers; instead some parts (branches) are preferred. The branching depends on the bounding:

(b) It is required that for any subbox $Y \subseteq X$ a lower bound for f over Y is known or computable.

The process of fitting together the branching and bounding is discussed later. Here we mention that we use interval arithmetic for the bounding since it has the tools for obtaining the necessary bounds almost automatically.

The 3 algorithms depend on:

(i) *Properties of f.* Principally, the algorithms work and converge even if f is not continuous. If f is continuous then it is easier to find sound termination criteria. The number of global minimum points

Introduction

can be unbounded, but f has to be bounded from below in order to have a global minimum. Lipschitz conditions, differentiability, or smoothness properties are not needed but the numerical processing is facilitated and the convergence speed may improve if they are present.

(ii) *Inclusion functions.* These are the interval valued functions which enable us to include the whole continuum of function values of f (and thus f^*) or, if desired, certain parts of it in intervals (bounding). Construction and use of inclusion functions cause no problems. When a computer system is available equipped with interval arithmetic software then the construction and arithmetic manipulation of inclusion functions run completely automatically and, further, the lower bounds computed are guaranteed lower bounds since the rounding errors are kept under control automatically. Majorization resp. minimization methods based on the knowledge of Lipschitz constants are closest tools to the inclusion function tool. Nevertheless, inclusion function methods are preferred since it is nearly always possible to find appropriate inclusion functions, even if they do not satisfy a Lipschitz condition or are not continuous.

(iii) *A subdivision (bisection) strategy* of the domain X. The inclusion function brings about guaranteed inclusions of f^*. The subdivision strategy implies that the inclusion function is evaluated only where it is really necessary (branching) so that the computational effort is kept as low as possible. The computation can be terminated when the inclusions of f^* are sufficiently small.

An effective procedure for a global unconstrained optimization problem will consist of

(a) the basic steps,

(b) the accelerating devices.

The basic steps are responsible for getting the solution of the problem or, at least, an approximate solution when the computation is done on a computer. The accelerating devices aim to get the solution or its approximation as fast as possible.

Our presentation of the 3 algorithms will only consist of the basic steps for two reasons: First, the convergence properties are then transparent and they can therefore easily be proven. Second, the basic steps require minimum assumptions for f. If, however, the user wishes to speed up the computation and if f has the properties required for speeding up the computations then he may add accelerating devices. These are described at the end of this chapter.

The organization of this chapter is as follows:

First, the Moore-Skelboe algorithm is presented. Its properties are discussed in depth since the remaining two algorithms can be seen as modifications of the above algorithm. Items like the existence of minimizers, termination criteria, influence of rounding errors, convergence properties of the algorithms, etc. are therefore investigated extensively in connection with the Moore-Skelboe algorithm since these items are basically valid for all 3 algorithms. When dealing with the properties of Ichida-Fujii's and of Hansen's algorithm it is therefore only necessary to add a few supplementary remarks. Two sections with numerical examples demonstrate that the interval methods are far from being mere theory. Some examples are included which are unsolvable with any other optimization procedure we know about. Finally, the acceleration devices are developed, and it is shown how they can be combined with the 3 prototype algorithms.

3.2 The Moore-Skelboe Algorithm

This algorithm aims to determine f^*, the global minimum of the unconstrained optimization problem, (3.1), that is, to minimize f over the box $X \in \mathbf{I}^m$. Some global minimizers can be located. An error analysis is possible only for the approximation of f.

The algorithm is credited to Moore and Skelboe because Moore (1966) was probably the first to discover that interval arithmetic is an excellent tool for computing the range of a function over a box X, which is almost the same as to compute f^*. Skelboe (1974) combined some of Moore's ideas with the branch and bound principle. The resulting algorithm was again improved by Moore (1976).

The input data for the algorithm are the dimension m of the prob-

The Moore-Skelboe Algorithm

lem (number of variables of f), the box $X \in \mathbf{I}^m$ and an inclusion function $F : \mathbf{I}(X) \to \mathbf{I}$ of f. Termination criteria will be discussed later.

The algorithm works by splitting up the domain X into subboxes of not necessarily the same size step by step. The search for f^* is done in these subboxes, but not uniformly which would be too expensive. Here the *branching principle* is involved: At each iteration the search is continued in the box Y where f has the smallest lower bound y (*bounding principle*), since the chances are best for finding f^* in this box. The lower bounds of f over subboxes Z are determined from the inclusion function $F(Z)$ such that $z = \mathrm{lb}F(Z)$ is a lower bound of f over Z.

ALGORITHM 1 *(Moore-Skelboe)*

1. *Set $Y := X$.*
2. *Calculate $F(Y)$.*
3. *Set $y := \min F(Y)$.*
4. *Initialize list $\mathbf{L} = ((Y, y))$.*
5. *Choose a coordinate direction k parallel to which $Y = Y_1 \times \ldots \times Y_m$ has an edge of maximum length, i.e. $k \in \{i : w(Y) = w(Y_i)\}$.*
6. *Bisect Y normal to direction k obtaining boxes V_1, V_2 such that $Y = V_1 \cup V_2$.*
7. *Calculate $F(V_1), F(V_2)$.*
8. *Set $v_i := \mathrm{lb}F(V_i)$ for $i = 1, 2$.*
9. *Remove (Y, y) from the list $\mathbf{L}$.*
10. *Enter the pairs (V_1, v_1) and (V_2, v_2) into the list such that the second members of all pairs of the list do not decrease.*
11. *Denote the first pair of the list by (Y, y).*
12. *If termination criteria hold, then go to 14.*
13. *Go to 5.*
14. *End*

This algorithm initializes a list $\mathbf{L} = \mathbf{L}_1$ consisting of one pair (Y, y); see Step 4. The list is then modified and enlarged at each iteration; see Steps 9 and 10. At the n-th iteration a list $\mathbf{L} = \mathbf{L}_n$ consisting of n pairs is present,

$$\mathbf{L}_n = ((Z_{n1}, z_{n1}), \ldots, (Z_{nn}, z_{nn})) \text{ where } z_{nk} = \min F(Z_{nk}).$$

The *leading pair* of the list $\mathbf{L}_n$ will be denoted by

$$(Y_n, y_n) = (Z_{n1}, z_{n1}).$$

The boxes Y_n are called the *leading boxes* of the algorithm.

Assume, for the moment, that the termination criteria of Step 12 are not satisfied during the whole computation such that Algorithm 1 will not stop. In this case an infinite sequence of leading pairs $(Y_n, y_n)_{n=1}^{\infty}$ is produced, and the convergence properties of Algorithm 1 can be discussed. The set of accumulation points of the sequences (Y_n) and (y_n) is then taken as solution set of Algorithm 1.

A point x is an *accumulation point* of the sequence (Y_n) iff there exist points $\eta_n \in Y_n$ for each n such that x is an accumulation point of the sequence (η_n). We can use the Hausdorff-metric, and a weakened version of it, in order to have a measure for the distance from the boxes Y_n to X^*, the set of global minimum points available. Let A, B be compact subsets of $\mathbf{R}^m$ and $x \in \mathbf{R}^m$; then we define

$$\begin{aligned} d_o(x, B) &= \min_{b \in B} \| x - b \|_2, \\ d_o(A, B) &= \max_{a \in A} d_o(a, B), \\ d(A, B) &= \max\{d_o(A, B), d_o(B, A)\}. \end{aligned}$$

Clearly, $d_o(x, B)$ is the distance from x to the nearest point of B, and, roughly speaking, $d_o(A, B)$ is the distance from the farthest point of A to the nearest point of B. Therefore $d_o(A, B) \neq d_o(B, A)$, in general. If, for example, $A \subseteq B$ then $d_o(A, B) = 0$, but $d_o(B, A) = 0$ need not hold in this case. The Hausdorff-distance d is a metric for the set of compact subsets of $\mathbf{R}^m$. Thus, $d(A, B) = 0$ only if $A = B$. This shows that convergence of boxes Y_n to a set B via d means that Y_n converges to B *with respect to the shape* of B, which is a requirement that is too strong in many cases. On the contrary, $d_o(Y_n, B) \to 0$ as

The Moore-Skelboe Algorithm

$n \to \infty$ means that the farthest point of Y_n from B tends to B. This is a realistic way to describe the kind of convergence we are concerned with in Alg. 1. We introduce the notation

$$Y_n \to B$$

only for the case $d(Y_n, B) \to 0$.

If B consists of just one point, say $B = \{x\}$, then $Y_n \to B$ or $Y_n \to z$ means that $\eta_n \to z$ for any sequence (η_n) with $\eta_n \in Y_n$. Therefore if $Y_n = [a_n, b_n]$ are intervals ($m = 1$), then $Y_n \to z$ is equivalent to $a_n \to z$ and $b_n \to z$.

Three kinds of results can be expected.

1. If *any* inclusion function F for f is the parameter of Alg. 1 then the sequence of leading pairs, $((Y_n, y_n))_{n=1}^{\infty}$, satisfies $f^* \in F(Y_n)$, i.e., $y_n \leq f^*$ for any n.

2. If F satisfies an additional property (cf. (3.12) or (3.13)) then the convergence of the algorithm is guaranteed in the following sense:

$$F(Y_n) \to f^* \quad \text{as } n \to \infty \text{ if } f \text{ is continuous,}$$
$$y_n \to f^* \quad \text{as } n \to \infty \text{ if } f \text{ is discontinuous.}$$

    In both cases, estimates of the error are automatically produced.

3. Each accumulation point of (Y_n) is a global minimizer.

These results are discussed in the sequel.

THEOREM 1 *Let $X \subseteq \mathbf{R}^m$ be a box, let $f : X \to \mathbf{R}$ be any function, and let $F : \mathbf{I}(X) \to \mathbf{I}$ be an inclusion function for f. Then the global "minimum" $f^* = \inf(\Box f(X))$ exists and*

$$f^* \in F(Y)$$

holds for all leading boxes Y of the algorithm.

Proof. The range of f over X is included in the interval $F(X)$ from the assumption that F exists. Therefore, $\inf(\Box f(x))$ exists in **R**. The remainder of the theorem follows immediately from the following formulas which hold for each list L_n:

$$X = \bigcup_{i=1}^{n} Z_{ni}, \qquad (3.2)$$

$$\Box f(X) = \bigcup_{i=1}^{n} \Box f(Z_{ni}) \subseteq \bigcup_{i=1}^{n} F(Z_{ni}), \qquad (3.3)$$

$$y \leq f^* \leq \text{ub} F(Y). \Box \qquad (3.4)$$

The relations (3.2), (3.3) show that the algorithm has similarities with known sequential strategies based on the generation of minorants which can be found via the knowledge of Lipschitz constants, cf. Evtushenko (1971), Shubert (1971), Mladineo (1986), etc.

Finally, two technical points should be mentioned: First, if f is monotone in some variable or if some variable occurs only once in the expression for F, then no subdivisions are necessary in the related coordinate direction, cf. Nickel (1971), Skelboe (1974), Ratschek-Rokne (1984).

Second, if a reduction of the list is desired, for instance to avoid capacity problems when storing the list in a computer, a cut-off process can be incorporated. This means that if the list L_n is processed and if

$$\max F(Z_{n1}) < z_{nj} \text{ for some } j \leq n,$$

then all pairs

$$(Z_{ni}, z_{ni}) \text{ for } i = j, \ldots, n$$

can be discarded from the list. Such a procedure is justified, since, for $i = j, \ldots, n$, the box Z_{ni} does not influence the search for f^*. This test is related to the so-called *midpoint test*. As we will see it is a basic step of Ichida-Fujii's as well as of Hansen's algorithms where the test influences the solution sets of the algorithms. The principle of this test is also used in non-interval optimization algorithms.

3.3 Termination, Approximation Errors, Rounding Errors

In order to provide termination criteria, one has to distinguish two cases: The criteria have to terminate the execution of the algorithm when

(i) exact computation is assumed (idealized case),

(ii) numerical computation is assumed (real case).

Both criteria are important for our considerations. The first criterion is responsible for the connection to the theory of the algorithm and the second criterion is responsible for the approximation of the first such that the connections with the theory are not lost. We start with a discussion of the criteria for the Moore-Skelboe Algorithm when exact computation is assumed.

A reasonable criterion is

$$\text{"If } w(F(Y_n)) < \epsilon \text{ then terminate"} \qquad (3.5)$$

where $\epsilon > 0$. Since $f^* \in F(Y_n)$ the number $w(F(Y_n))$ is an *upper bound for the absolute error* when f^* is approximated by $F(Y_n)$ or any value $\eta_n \in F(Y_n)$, for example, $\eta_n := y_n$.

Practically one obtains better results when the criterion

$$\text{"If } f_n - y_n < \epsilon \text{ then terminate"} \qquad (3.6)$$

is used. Here f_n is the smallest function value (of f) which has been computed up to the n-th iteration. Since $y_n \leq f^* \leq f_n$, we again have an upper estimate of the approximation error $\mid y_n - f^* \mid$ or $\mid f_n - f^* \mid$ by $f_n - y_n$.

The function values f_n are in many cases available when the inclusion function values $F(Y)$ are computed. For instance, if $F(Y)$ is obtained via the meanvalue form,

$$F(Y) = f(c) + (Y - c)^T F'(Y)$$

where $c = \text{mid } Y$ and where F' is an inclusion of the gradient or the generalized gradient, then $f(c)$ is such a function value. If really no function value is available then, in general,

$$f_n := \min_{i=1,\ldots,n} \text{ub} F(Y_n)$$

will be the smallest upper bound of f^* known so far and it may be used in (3.6).

One also can obtain sharper results if

$$\hat{y}_n := \max\{y_1, \ldots, y_n\}$$

is used instead of y_n. Clearly $f_n - \hat{y}_n$ is smaller than $f_n - y_n$ and we also have $\hat{y}_n \leq f^*$. In the sequel, we do not mention this possibility since its use is a programming skill rather than an important step of Alg. 1 or other algorithms which will be developed in this book.

We can also estimate the relative approximation error. If the computation is terminated by (3.5) then the relative error when f^* is approximated by $F(Y_n)$ or by any value $\eta_n \in F(Y_n)$ is, maximally,

$$\frac{w(F(Y_n))}{\min(|y_n|, |\text{ub}F(Y_n)|)} \quad \text{if } 0 \notin F(Y_n). \tag{3.7}$$

If the termination is caused by (3.6) then the relative error of the approximation of f^* by y_n or f_n is, maximally,

$$\frac{f_n - y_n}{\min(|y_n|, |f_n|)} \quad \text{if } 0 \notin [y_n, f_n]. \tag{3.8}$$

Using (3.5) or (3.6) the termination is guaranteed when $F(Y_n) \to f^+$. Conditions for the convergence properties of this sequence are given in the next section. If convergence is not given or if rounding errors prevent the convergence during the numerical computation, then other criteria have to be incorporated, or additionally used, for security. Two of them are, for instance,

"If $w(Y_n) < \epsilon_o$ then terminate"

for some $\epsilon_o > 0$. This criterion always works since we always have $w(Y_n) \to 0$ as we will see in the next section. Or,

"If $n \geq n_o$ then terminate."

The error estimations mentioned above remain valid also under the last two criteria.

So far we have ignored the effect of rounding errors in order to be able to study the theoretic properties of Alg. 1. If Alg. 1 is implemented on a computer then the assertions of Theorem 1 may be wrong. The inclusion property of f^* need no longer be preserved and convergence properties as they will be presented in the next section are disturbed or destroyed due to rounding errors. Then instead of the intended values $F(Y)$, so-called numerical values $\tilde{F}(Y)$ are delivered which approximate $F(Y)$.

If, however, *machine interval arithmetic* is implemented on the computer, then one has the rounding errors automatically under control; see Sec. 2.4. In addition to other advantages, a machine interval arithmetic causes the inclusion

$$F(Y) \subseteq \tilde{F}(Y) = (F(Y))_M$$

such that the global minimum remains included,

$$f^* \in \tilde{F}(Y)$$

if Y is a leading box. If $\tilde{y} = \mathrm{lb}\tilde{F}(Y)$ and if Y is a leading box then

$$\tilde{y} \leq \mathrm{lb}F(Y) \leq f^*;$$

thus the numerical value $\tilde{y}$ of $y = \mathrm{lb}F(Y)$ is still a lower bound for f^*. The termination condition $w(F(Y)) < \epsilon$, cf. (3.5), is then numerically realized as the condition

$$w(\tilde{F}(Y)) < \epsilon. \tag{3.9}$$

If (3.9) is satisfied then it also follows that $w(F(Y)) < \epsilon$ such that the termination of the algorithm remains correct, and

$$w(\tilde{F}(Y)) < \epsilon$$

is an absolute error bound of $f^* - y$ such that Theorem 1 is still valid. It is possible that in a certain phase of the computation the rounding errors overwhelm the decrease of $w(F(Y))$, that is, $w(\tilde{F}(Y))$ does not tend to 0 even if $w(F(Y)) \to 0$. Since such cases cannot be excluded, additional termination criteria are required as mentioned above. The following two security criteria are also known:

A widely used condition for termination was proposed in Moore (1966). That is, the termination of Alg. 1 is initiated when

$$Y_{n+1} \subseteq Y_n \text{ does not imply } \tilde{F}(Y_{n+1}) \subseteq \tilde{F}(Y_n) \qquad (3.10)$$

where Y_n and Y_{n+1} are the leading boxes of L_n and L_{n+1}, respectively. This criterion, however, presupposes that the inclusion function is isotone. (F is called *isotone* if $Y \subseteq Z$ implies $F(Y) \subseteq F(Z)$.) For this reason the investigation and construction of isotone interval functions is an important branch of interval analysis.

If f is discontinuous the graph of f may contain jumps and the termination criterion $w(F(Y_n)) < \epsilon$ may not work. The criterion (3.10) is then recommended as an additional termination criterion as well. If F is inclusion isotone, then a few occurrences of (3.10) without interruption will in general suggest that the result cannot be improved. There is a similar criterion that causes termination if the sequence $(y_n)_{n=1}^{\infty}$ does not increase during some steps of the algorithm.

Some authors recommend stopping the computation when the condition

$$Y_{n+1} \subseteq Y_n, Y_{n+1} \neq Y_n \text{ implies } \tilde{F}(Y_{n+1}) = \tilde{F}(Y_n) \qquad (3.11)$$

occurs during some steps of the algorithm. A class of functions for which (3.11) provides an optimal termination criterion is described in Asaithambi-Shen-Moore (1982). When applying (3.11) to other functions one should not stop too early: One can construct examples where at first (3.11) holds arbitrarily many times without interruptions, but an improvement is nevertheless possible. For example, if

$$f(x) = \sin(2\pi x) + \cos(2\pi x), \quad F(Y) = \square\sin(2\pi Y) + \square\cos(2\pi Y),$$

and $X = [0, 2^{10}]$, then $\square\sin(2\pi Y) = \square\cos(2\pi Y) = [-1, 1]$ if $w(Y) \geq 1$. Thus, we will get $F(Y_n) = [-1, 1] + [-1, 1] = [-2, 2]$ and $w(F(Y_n)) = 4$ for $n = 1, \ldots, 2^{11} - 1$, when applying Alg. 1 to X. Only then is there a chance that a decreasing $w(F(Y_n))$ is computed.

3.4 Convergence Conditions for the Moore-Skelboe Algorithm

It was shown in Sec. 3.3 that Alg. 1 produces intervals $F(Y_n)$ that contain the global minimum f^*, such that $y_n = \mathrm{lb} F(Y_n)$ is a lower bound for f^* and $w(F(Y_n))$ is a measure for the quality of the approximation. In this section we shall first discuss conditions under which Alg. 1 converges, that is, $w(F(Y_n))$ converges to 0 or Y_n converges to f^* as $n \to \infty$, and then give some hints about the convergence speed. The results of this section are mainly due to Ratschek (1985a) and Moore-Ratschek (1987).

Let us first state a key lemma for further investigations. It expresses that for any sequence of intervals generated by Alg. 1, $(Y_n)_{n=1}^\infty$, the corresponding sequence of widths $(w(Y_n))_{n=1}^\infty$ converges to 0. This property may seem to be self-evident at first glance. However, the proof is useful for three reasons. Firstly, there are very similar algorithms that do not have this property, for example algorithms which use cyclic bisection: see Moore (1979), Asaithambi-Shen-Moore (1982). Secondly, the sequence $(Y_n)_{n=1}^\infty$ need not converge. Thirdly, we may reflect on the possible occurrence of sequences $(Y_n)_{n=1}^\infty$ where boxes of length $w(X)/2^v$ for $v = 1, 2, \ldots$ may arise arbitrarily late. Such a sequence arises for example when for each $v = 1, 2, \ldots$ a box Y_n exists with $n \geq 10^v$ and $w(Y_n) = 1/2^v$.

LEMMA 1 *Let* $(Y_n)_{n=1}^\infty$ *be a sequence generated by Alg. 1. Then* $w(Y_n) \to 0$ *as* $n \to \infty$.

Proof. Let $X \in \mathbf{I}^m$ be the basis box. Then the construction of the sequence of leading boxes depends on F and on the geometry of X, that is, the length of the edges of X. If X is fixed but another inclusion function F is chosen then another sequence of boxes emerges. Let $\mathcal{S}$ be the collection of all boxes that occur if the algorithm is applied to X and all possible inclusion functions F. Then $\mathcal{S}$ contains only finitely many boxes whose width is larger than any given $\epsilon > 0$. This shall be shown now.

Let the coordinate directions be numbered in such a way that $w_1 \geq w_2 \geq \ldots \geq w_m$ where $w_i = w(X_i)$ and $X = X_1 \times \cdots \times X_m$.

Then no box of $\mathcal{S}$ has a width larger than w_1. Estimating the number of boxes of $\mathcal{S}$ that satisfy $w_1/2 < w(Y) \leq w_1$ leads to an upper bound, α, for this number, which is $\alpha = 3^m$. The number is at most m if $w_1 > w_2 > \cdots > w_m$ and is equal to α if all the w_i are equal. Further, there are at most α boxes of $\mathcal{S}$ with width $w_1/2$. Since this estimate is independent of the numerical value of w_1 we can apply it again to at most α boxes with the edge length $w'_i = w_i/2$ and $w'_1 \geq w'_2 \geq \cdots \geq w'_m$. Accordingly, $\mathcal{S}$ contains at most α^2 boxes Y of width $w_1/4 < w(Y) \leq w_1/2$. Continuing this process we see that for each k, the number of boxes of $\mathcal{S}$ with a width larger than $w_1/2^k$ is at most $\alpha + \alpha^2 + \cdots + \alpha^k$, i.e. finite. Since $(w_1/2^k)_{k=1}^{\infty}$ is a null sequence it follows that, given any $\epsilon > 0$, there are only finitely many boxes Y of $\mathcal{S}$ with $w(Y) \geq \epsilon$.

The lemma will have been proven when, given an $\epsilon > 0$, there exists a number n_0 such that $w(Y_n) < \epsilon$ for any $n \geq n_0$. Let us assume that there exists an $\epsilon > 0$ such that for any n_0 an $n \geq n_0$ exists with $w(Y_n) \geq \epsilon$. This means that a subsequence $(Y_{k_v})_{v=0}^{\infty}$ of $(Y_n)_{n=1}^{\infty}$ exists such that $w(Y_{k_v}) \geq \epsilon$ for $v = 0, 1, \ldots$. Since $Y_{k_v} \in \mathcal{S}$ and $\mathcal{S}$ contains only finitely many boxes Y with $w(Y) \geq \epsilon$ we have a contradiction. □

With this lemma in mind the convergence properties of Alg. 1 can be shown almost without assumptions. One has only to choose an inclusion function F of f with the property

$$w(F(Y)) \to 0 \text{ as } w(Y) \to 0. \tag{3.12}$$

This is a very natural condition and to find such inclusions causes no problems at all, not even if f is not explicitly given, which is for example the case if f is recursively defined or defined via a numerical algorithm. Let $((Y_n, y_n))_{n=1}^{\infty}$ be again the sequence of pairs generated by Alg. 1. Then we have the following result:

THEOREM 2 *If the inclusion function F for f satisfies (3.12), then the sequence $(F(Y_n))_{n=1}^{\infty}$ converges to the global minimum f^*, i.e., the sequence $(y_n)_{n=1}^{\infty}$ converges to f^* from below.*

Proof. By the lemma, $w(Y_n) \to 0$ as $n \to \infty$. Thus, by (3.12), also $w(F(Y_n)) \to 0$ as $n \to \infty$. Since $f^* \in F(Y_n)$ for all n by Theorem 1, $F(Y_n)$ and also $y_n = \text{lb} F(Y_n)$ tend to f^*. □

Contrary to Theorem 1, Theorem 2 holds only for continuous functions f. The continuity of f is implied by the assumption (3.12). The convergence property of Alg. 1, however, is not lost if the continuity of f is dropped, but it is difficult to find a realistic termination criterion. If f is not continuous, then (3.12) has to be replaced by

$$w(F(Y)) - w(\Box f(Y)) \to 0 \text{ as } w(Y) \to 0. \quad (3.13)$$

Since $\Box f(Y)$ need not be a compact interval we use $w(\Box f(Y))$ to denote the width of the interval hull of $\Box f(Y)$, which is the smallest compact interval that contains $\Box f(Y)$. It is obvious that (3.12) and (3.13) are equivalent if f is continuous. The following assertion describes the convergence properties of Alg. 1 applied to discontinuous functions.

THEOREM 3 *If the inclusion function F for f satisfies (3.13) then the sequence $(y_n)_{n=1}^{\infty}$ converges to f^* from below.*

Proof. By the lemma, the sequence $(w(Y_n))_{n=1}^{\infty}$ converges to 0. Thus, by (3.13), $w(F(Y_n)) - w(\Box f(Y_n)) \to 0$ as $n \to \infty$. From $\Box f(Y_n) \subseteq f(Y_n)$ it follows that $\inf \Box f(Y_n) - y_n \to 0$ as $n \to \infty$. Together with $y_n \le f^* \le \inf \Box f(Y_n)$, the assertion follows. $\Box$

Theorem 3 says that $(y_n)_{n=1}^{\infty}$ converges to f^*. However, the sequence $(F(Y_n))_{n=1}^{\infty}$ need not converge to f^* as is the case in Theorem 2 since it is possible that the intervals $F(Y_n)$ contain a jump of f.

Although the sequence $(w(Y_n))_{n=1}^{\infty}$ converges to zero, the sequence of leading intervals, $(Y_n)_{n=1}^{\infty}$, need not converge in general. If x^* is the only global minimum point of f in X then $\lim Y_n = x^*$. If there are several global minimum points the sequence $(Y_n)_{n=1}^{\infty}$ can converge to one of them but in most cases there will be subsequences that converge to some of these minimum points.

Let us now turn to the behaviour of Alg. 1 with respect to X^*, the set of global minimizers of f over X. The set of accumulation points of the sequence (Y_n) is denoted by A. More precisely, a point x belongs to A, if x is an accumulation point of some sequence (η_n) with $\eta_n \in Y_n$ for all n. Obviously, X^* and A are compact subsets of X.

THEOREM 4 *Let f be continuous and let $w(F(Y)) \to 0$ as $w(Y) \to 0$. Then*

(i) $A \neq \emptyset$ *(the set of accumulation points is non-empty)*,

(ii) $A \subseteq X^*$ *(that is, each accumulation point is a global minimum point)*,

(iii) $d_o(Y_n, A) \to 0$ *as $n \to \infty$*.

Proof. (i) Let $(\eta_n)_{n=1}^\infty$ be a sequence such that $\eta_n \in Y_n$. Since $Y_n \subseteq X$ and X is compact, that sequence has an accumulation point because of the Bolzano-Weierstrass theorem.

(ii) Let $\tilde{x}$ be an accumulation point of the sequence $(\eta_n)_{n=1}^\infty$ with $\eta_n \in Y_n$. Then there exists a subsequence $(\eta_{k_n})_{n=1}^\infty$ which converges to $\tilde{x}$. Since $\eta_{k_n} \in Y_{k_n}$ and since $w(Y_n) \to 0$ as $n \to \infty$ (see Lemma 1), the corresponding subsequence (Y_{k_n}) also tends to $\tilde{x}$. Since f is continuous, we get $\Box f(Y_{k_n}) \to f(\tilde{x})$. Since $F(Y_n)$ tends to the global minimum, f^*, see Theorem 2, and since $\Box f(Y_{k_n}) \subseteq F(Y_{k_n})$, it also follows that $\Box f(Y_{k_n}) \to f^*$. Therefore, $f(\tilde{x}) = f^*$, that is, $\tilde{x} \in X^*$.

(iii) Since $w(Y_n) \to 0$, the assertion is equivalent to $d_o(c_n, A) \to 0$ as $n \to \infty$ where $c_n = \mathrm{mid}\,(Y_n)$, that is the midpoint of Y_n. In order to get a proof by contradiction, we assume that $d_o(c_n, A)$ does not converge to 0. So an $\epsilon > 0$ and a subsequence $(c_{k_n})_{n=1}^\infty$ of $(c_n)_{n=1}^\infty$ exists such that $d_o(c_{k_n}, A) = \min_{\tilde{x} \in A} \| c_{k_n} - \tilde{x} \|_2 \geq \epsilon$ for all n. But $(c_{k_n})_{n=1}^\infty$ is a sequence in a compact set, X, and so has an accumulation point $\tilde{\tilde{x}} \in A$. $\Box$

Remarks. (1) If $\tilde{x}$ is the *only* global minimizer, then the sequence (Y_n) *converges* to $\tilde{x}$, since $A = X^* = \{\tilde{x}\}$ by (i) and (ii) of Theorem 4 in this case.

(2) The sets A of accumulation points and X^* of global minimizers can be assumed to be equal from a *practical* point of view. The reason is that a point $x^* \in X^*$ does not belong to A if and only if there exists

an index n and a *nonleading* pair (Z_{ni}, z_{ni}) in L_n such that

$$x^* \in Z_{ni} \text{ and } f^* = z_{ni}$$

and further, that $Z_{ni} = Z_{mi_m}$ for all $m > n$ and an appropriate position i_m in L_m. This means that the box Z_{ni} will never be further subdivided. These three conditions will practically almost never occur (at least, if trivial functions and sophisticated examples are excluded). The conditions mean that there exist other global minimum points $\tilde{x}$ which determine the leading pairs (Y_n, y_n) and where $y_n < f^* = z_{ni}$ such that z_{ni} can never be leading. Nevertheless, the Lebesgue measure $\lambda(X^* \setminus A)$ may be positive.

(3) Assertion (iii) of the theorem cannot be replaced by $d(Y_n, A) \to 0$. This would imply that $d_o(A, Y_n) \to 0$ which is only the case if A is a single point.

(4) No termination criteria exist for Algorithm 1 which are based on the distance between the global minimizers and their approximations by the leading intervals Y_n.

(5) A result similar to assertion (ii) of Theorem 4 was derived by Benson (1982) who studied a prototype branch and bound algorithm under comparable assumptions.

3.5 Numerical Examples

The Moore-Skelboe Algorithm and its properties are tested first on two "bad" functions in order to demonstrate the superfluousness of frequently used assumptions for its application. Finally the effect of using different inclusion functions dealing with the six hump camel back function will be illustrated. The interested reader will find several further numerical results which are concerned with differentiable functions in Skelboe (1974), Asaithambi-Shen-Moore (1982).

The following examples which can be found in Ratschek (1985a) were calculated on an Apple IIe microcomputer equipped with a PASCAL-SC software system. The symbols which are used in the sequel mean:

$X = X_1 \times X_2$	Basic domain (Step 1 of the algorithm).
ϵ	Intended absolute accuracy (Termination criterion (3.6)).
N	Number of function evaluations of F till termination.
(Y^*, y^*)	Leading pair when terminating, $Y^* = Y_1^* \times Y_2^*$.
y^*	As just defined is a lower bound of the global minimum, f^*.
ϵ_0	Attained absolute accuracy which has been computed as minimum of the values $f(x) - y^*$ for the 4 corners x of Y^*. Thus, $f^* - y^* \leq \epsilon_0$.
$E \pm k$	$10^{\pm k}$.

If the functions are not continuous the criterion used ((3.6)) does not guarantee a termination. Thus we added an emergency termination criterion as follows where the updated values $\hat{y}_n = \max_{i=1,\ldots,n} y_i$, cf. Sec. 3.3, are used:

"If $\hat{y}_n = \hat{y}_{n+20}$ then terminate."

This means that the computation stops if the updated lower bound has not been improved within 20 iterations.

Example 1. $f(x_1, x_2) = \begin{cases} x_1 \sin(1/x_1) + x_2 \mid x_1 \mid & \text{if } x_1 \neq 0, \\ 0 & \text{if } x_1 = 0. \end{cases}$

We are mainly interested in the case $x_2 \geq 1$ or where x_2 belongs to a neighborhood of 1, since the function

$$f(x_1, 1) = x_1 \sin(1/x_1) + \mid x_1 \mid$$

has infinitely many global minimum points, and 0 which is itself a global minimum point is an accumulation point of the remaining global minimum points. The global minimum for $f(x_1, 1)$ is 0. If x_2 is close to 1 then there are infinitely many local minimum points and most of the corresponding minimum values are close to zero.

Numerical Examples

We used the following inclusion function, defined for intervals $Y_1, Y_2 \in \mathbf{I}$,

$$\begin{aligned}
F(Y_1, Y_2) &= Y_1(\mathrm{ISIN}(1/Y_1) + Y_2) && \text{if } Y_1 > 10^{-90}, \\
&= Y_1(\mathrm{ISIN}(1/Y_1) - Y_2) && \text{if } Y_1 < -10^{-90}, \\
&= |Y_1|((Y_2 - 1) \vee (Y_2 + 1)) && \text{otherwise.}
\end{aligned}$$

ISIN denotes the inclusion function of the sine function in PASCAL-SC; see Sec. 2.6.

$$|Y_1| = \max\{|x_1| : x_1 \in Y_1\}$$

is the *absolute value* of an interval Y, and $U \vee V$ denotes the *interval hull* of two intervals or points, U and V, that is, the smallest interval containing both U and V. The absolute value of an interval and the convex hull are pre-declared functions in PASCAL-SC.

We applied the algorithm to F on four input boxes X. The results are shown in Tables 3.1 and 3.2.

X_1	[0, 2]	[−100, 100]
X_2	[0.99999, 2]	[−100, 100]
ϵ	$1E-6$	$1E-6$
N	72	68
ϵ_0	$5E-10$	$1E-8$
y^*	$-2.12211\ 609E-6$	$-9.99900\ 1E+3$
Y_1^*	$[2.1220, 2.1222]E-1$	$[0.99609, 1.0E+2]$
Y_2^*	$[9.9999, 9.99994]E-1$	$[-1.0, -0.99609]E+2$

Table 3.1.

X_1	[0, 1]	[0, 1]
X_2	[1, 1.00000001]	[1, 100]
ϵ	$1E-6$	$1E-6$
N	240	40
ϵ_0	$1.3E-7$	0.0
y^*	0.0	0.0
Y_1^*	$[2.12158, 2.12281]E-1$	[0.0, 0.25]
Y_2^*	[1.0, 1.00000 001]	[1.0, 1.38672]

Table 3.2.

The last test in Table 3.2 shows a number N of inclusion function evaluations of only 40 where the minimum $f^* = 0$ is determined exactly because of

$$f^* \in [y^*, y^* + \epsilon_0] = [0, 0].$$

This result may seem surprising; however, it is not since it was caused by the emergency termination criterion described previous to this example, i.e. the lower bounds $\tilde{y}_k = 0$ did not change during a certain number of iterations.

Example 2. $f(x_1, x_2) = 2 + x_1 - [10x_1]/10 + x_2 - [10x_2]/10 - \cos\sqrt{([10x_1]/10^2 + ([10x_2]/10)^2}$ if $x_1, x_2 \geq 0$, where $[r]$ denotes the largest integer smaller than or equal to the real number r. $[r]$ can be programmed via the standard function trunc in a programming language if $r \geq 0$; for example, $[3.5] = 3$.

The function f is discontinuous if $x_1 = 0.1, 0.2, 0.3$, etc., or if $x_2 = 0.1, 0.2, 0.3$, etc. This means that there are orbits of discontinuous points in both coordinate directions. The points $(x_1, x_2) = (k/10, l/10)$ are local minimum points if $k, l = 1, 2, 3$, etc. The only global minimum point is $(0, 0)$ and the global minimum of f is $f^* = 1$.

Numerical Examples

We used the following inclusion functions, defined for intervals $Y_1 = [y_{11}, y_{12}] \in I$ and $Y_2 \in I$, $y_{11} \geq 0$, $\min Y_2 \geq 0$,

$$G(Y_1) = Y_1 \quad \text{if } [10y_{11}] = [10y_{12}],$$
$$= [0, 1/10] \quad \text{otherwise,}$$

$$H(Y_1) = y_{11}^2 \vee y_{12}^2,$$
$$F(Y_1, Y_2) = 2 + G(Y_1) + G(Y_2) -$$
$$\text{ICOS}(\text{ISQRT}\,(H(Y_1) + H(Y_2))/10).$$

G is an inclusion function for $x_1 - [10x_1]/10$ if $x_1 \geq 0$, and H is an inclusion function for $[10x_1]^2$. The functions ICOS, ISQRT are inclusion functions for $\cos$, $\sqrt{\ }$ which are predefined in PASCAL-SC. Thus, F is an inclusion function for f.

X_1	[0, 1]
X_2	[0, 1]
ϵ	$1E-6$
N	82
ϵ_0	$2E-11$
y^*	$9.99999\ 99999E-1$
Y_1^*	$[0.0, 3.90625]E-3$
Y_2^*	$[0.0, 3.90625]E-3$

Table 3.3.

Table 3.3 shows that Alg. 1 had no difficulties at all when applied to this example.

Example 3. The six hump camel back function is given by

$$f(x_1, x_2) = 4x_1^2 - 2.1x_1^4 + \frac{1}{3}x_1^6 + x_1 x_2 - 4x_2^2 + 4x_2^4.$$

Since f is a polynomial and thus infinitely often differentiable one ought to solve the problem using accelerating devices as discussed in the sequel. We nevertheless took f as an example, since we wish to compare two different inclusion functions for f. The first one is

$$F(Y_1, Y_2) = Y_1^2(4 + Y_1^2(-2.1 + \frac{1}{3}Y_1^2)) + 4Y_2^2(Y_2^2 - 1) + Y_1 Y_2$$

where $Y_1^2 = \{y_1^2 : y_1 \in Y_1\}$. We note that $Y_1^2 \neq Y_1 Y_1$ if $Y_1 Y_1$ is formed as the interval product of Y_1 by itself using the product rule; however, $Y_1^2 \subseteq Y_1 Y_1$ is always valid. The second inclusion function is

$$\begin{aligned} G(Y_1, Y_2) = &\; f(c_1, c_2) + f_1'(c_1, c_2)(Y_1 - c_1) + f_2'(c_1, c_2)(Y_2 - c_2) \\ &+ H_1(Y_1, Y_2)(Y_1 - c_1)^2/2 + H_2(Y_1, Y_2)(Y_2 - c_2)^2/2 \\ &+ (Y_1 - c_1)(Y_2 - c_2) \end{aligned}$$

where c_1 and c_2 denote the midpoints of the intervals Y_1 and Y_2, resp., f_1', f_2' are the components of f', and

$$\begin{aligned} H_1(Y_1, Y_2) &= 8 + Y_1^2(-25.2 + 10Y_1^2), \\ H_2(Y_1, Y_2) &= -8 + 48Y_2^2. \end{aligned}$$

G arises by developing f as a Taylor polynomial of second order and by replacing several parts with inclusion functions. Thus G is a Taylor-form of second order, cf. Sec. 2.7. In Table 3.4 the difference between F and G is first illustrated by evaluating them at a few arguments. Since G is quadratically convergent (see Sec. 2.7) one may expect better results with small arguments.

Y_1	$[-2.5, 2.5]$	$[-0.5, 0.5]$	$[0.99, 1.01]$	$[8.0, 9.0]E - 2$
Y_2	$[-1.5, 1.5]$	$[-0.5, 0.5]$	$[0.99, 1.01]$	$[-7.2, -7.1]E - 1$
$F(Y_1, Y_2)$	$[-70, 40]$	$[-1.25, 1.25]$	$[3.04, 3.43]$	$[-1.068, -0.995]$
$w[F(Y_1, Y_2)]$	110	2.5	0.39	$7.2E - 2$
$G(Y_1, Y_2)$	$[-480, 870]$	$[-1.25, 1.75]$	$[3.11, 3.36]$	$[-1.032, -1.031]$
$w[G(Y_1, Y_2)]$	1350	3.0	0.25	$1.2E - 3$

Table 3.4.

The following numerical test shows the different results when F and G were used to reach an accuracy of only 10^{-1}. Demanding an accuracy of 10^{-6} was too much for F; we had stack overflow at about 1800 evaluations of F because the list L became too long.

	F	G	G
X_1	$[-2.5, 2.5]$	$[-2.5, 2.5]$	$[-2.5, 2.5]$
X_2	$[-1.5, 1.5]$	$[-1.5, 1.5]$	$[-1.5, 1.5]$
ϵ	$1E-1$	$1E-1$	$1E-6$
N	1236	244	664
ϵ_0	$4.9E-2$	$4.2E-2$	$1.9E-7$
y^*	-1.0799	-1.0472	-1.031628585
Y_1^*	$[-0.977, 1.172]E-1$	$[0.0, 1.563]E-1$	$[8.972, 9.003]E-2$
Y_2^*	$[-7.266, -7.149]E-1$	$[-7.5, -6.563]E-1$	$[-7.1265, -7.1246]E-1$

Table 3.5.

3.6 Convergence Speed of the Moore-Skelboe Algorithm

We now begin to investigate the convergence speed of the Moore-Skelboe algorithm. It will be shown in this section that Alg. 1, i.e. the convergence order of the sequence (y_n) where $y_n = \text{lb} F(Y_n)$, can be as slow as possible although the inclusion functions used are of arbitrarily high order (for a definition see Sec. 2.7). In the next section, we concentrate on isotone inclusion functions (as defined in the sequel) and show that the convergence order of the sequence (y_n) is reasonable if Alg. 1 is applied to isotone inclusion functions.

Formulas for the convergence order are unrealistic in most of the cases since the formulas have to consider the worst possible case. Thus, investigations of convergence order are called "worst case analysis". Practically, the worst case will not occur - the more, since Alg. 1 and also the other algorithms described in this book will be combined with accelerating devices. These devices will not influence the worst case analysis of the convergence speed since one can always construct examples such that the devices do not apply.

If one compares our results with the convergence order of typical methods for solving nonlinear problems one has to keep in mind that these methods are *global* methods and that the typical methods offering superlinear or quadratic convergence are *local* methods. We do not know any other global method which does not depend exponentially on the dimension, which means that the error of the n-th approximate is of the form $\mathcal{O}(n^{-cm})$ where c is a constant and m the dimension. Thus, the relatively slow speed of the algorithms presented in our book is caused by the *global* approach and not by the use of interval tools. Furthermore in order to have at least the exponential order, one has to use isotone inclusion functions. This forms the content of this and the next section. We also learn in the next section that, practically, the average order is between polynomial and exponential order.

For technical, i.e. for arithmetic, reasons we start the counting of the lists generated by Alg. 1 by 0 (the formulas then look nicer) in this and the following section. Thus the starting list is L_0 consisting of $(Y_0, y_0) = (Z_{01}, z_{01})$. For this reason, the list L_n has not only n pairs, but $n+1$ pairs,

$$L_n = ((Z_{nv}, z_{nv}))_{v=1}^{n+1}$$

since the counting of the pairs which is done by the second subscript, v, starts - as in the former sections - by 1.

Before we start, the reader should be reminded that an inclusion function $F : \mathbf{I}(X) \to \mathbf{I}$ for $f : X \to \mathbf{R}$ is called of *order* (also: convergence order) $\alpha > 0$ if

$$w(F(Y)) - w(\Box f(Y)) = \mathcal{O}(w(Y)^\alpha) \text{ for all } Y \in \mathbf{I}(X),$$

that is, if there exists a constant c such that

$$w(F(Y)) - w(\Box f(Y)) \leq cw(Y)^\alpha \text{ for all } Y \in \mathbf{I}(X). \tag{3.14}$$

F is called isotone if $Y \subseteq Z$ implies $F(Y) \subseteq F(Z)$ for all $Y, Z \in \mathbf{I}(X)$.

Again, let (y_n) be the sequence of (leading) lower bounds if Alg. 1 is applied to X and F.

LEMMA 2 *If F is isotone the sequence (y_n) is monotonically increasing.*

Proof. Let us consider the list $\mathbf{L}_n = ((Z_{nv}, z_{nv}))_{v=1}^{n+1}$ with

$$y_n = z_{n1} \leq z_{n2} \leq \ldots \leq z_{n,n+1}.$$

The bisection of $Y_n = Z_{n1}$ with $Y_n = V_1 \cup V_2$ gives new lower bounds $v_1 = \text{lb}F(V_1)$ and $v_2 = \text{lb}F(V_2)$. Since F is isotone we get

$$y_n \leq v_1, \quad y_n \leq v_2.$$

From the way $\mathbf{L}_{n+1}$ is constructed the leading value y_{n+1} is the smallest value of z_{n2}, v_1 and v_2. Since these 3 values are larger than or equal to y_n, we also have $y_n \leq y_{n+1}$. □

We show that our problem can be transformed to a simpler problem. Let $X \in \mathbf{I}^m$ and $f : X \to \mathbf{R}$ again be given. The function f need not be continuous; however, we assume that f has a global minimum f^* in X. Let F be an inclusion function for f. Instead of considering f and F we will consider the constant function $h : X \to \mathbf{R}$ with $h(x) = f^*$ for all $x \in X$, and

$$\begin{aligned}H(Y) &= [\text{lb}F(Y), \text{lb}F(Y) + w(F(Y)) - w(\Box f(Y))] \text{ if } F(Y) \leq f^* \\ &= [f^*, f^* + w(F(Y)) - w(\Box f(Y))] \text{ otherwise.}\end{aligned}$$

Let $\mathbf{N}$ be the set of nonnegative integers.

LEMMA 3 *For any $c, \alpha > 0$ we have $w(F(Y)) - w(\Box f(Y)) \leq cw(Y)^\alpha$ for all $Y \in \mathbf{I}(X)$ iff $w(H(Y)) - w(\Box h(Y)) = w(H(Y)) \leq cw(Y)^\alpha$ for all $Y \in \mathbf{I}(X)$. That is, F and H have the same order.* □

LEMMA 4 *If F is isotone then F and H generate the same sequence of lower bounds, (y_n), when Alg. 1 is applied.*

Proof. Let $L_n = ((Z_{nv}, z_{nv}))_{v=1}^{n+1}$ with $Z_{n1} = Y_n, z_{n1} = y_n$ for $n \in \mathbf{N}$ be the lists generated by F and let $\tilde{L}_n = ((\tilde{Z}_{nv}, \tilde{z}_{nv}))_{v=1}^{n+1}$ with $\tilde{Z}_{n1} = \tilde{Y}_n, \tilde{z}_{n1} = \tilde{y}_n$ for all $n \in \mathbf{N}$ be the lists generated by H. Let γ_n be the number of pairs (Z_{ni}, z_{ni}) of L_n with $z_{ni} < f^*$. Let $\tilde{\gamma}_n$ be defined analogously.

Let us look at some properties of F and H. If $\gamma_k = 0$ for some k, that is, $y_k = f^*$, then $y_n = f^*$ and $\gamma_n = 0$ for all $n \geq k$. This is due to the monotonicity of the sequence (y_n) verified in Lemma 2. The corresponding inclusion function H need not be isotone. If, however, $\mathrm{lb}H(Z) = f^*$ for some box Z then $\mathrm{lb}H(V) = f^*$ for each subbox V of Z. This also results from the isotonicity of F. Therefore, if $\tilde{y}_k = f^*$ for some k then, considering the order of $\tilde{L}_k$, we get $\tilde{z}_{ki} = f^*$ for all i. Further, the bisection $\tilde{Y}_k = \tilde{V}_1 \cup \tilde{V}_2$ produces lower bounds $\tilde{v}_j = \mathrm{lb}H(\tilde{V}_j) = f^*$ for $j = 1, 2$ such that finally $\tilde{z}_{ni} = f^*$ for all $n \geq k$ and all i. That is, if $\tilde{\gamma}_k = 0$ for some k then $\gamma_n = 0$ for $n \geq k$.

We will now prove by induction on the list index n that

(i) $\gamma_n = \tilde{\gamma}_n$,

(ii) $(Z_{ni}, z_{ni}) = (\tilde{Z}_{ni}, \tilde{z}_{ni})$ for $i \leq \gamma_n$,

(iii) $y_n = \tilde{y}_n = f^*$ if $\gamma_n = 0$.

The lemma will then follow from these items. If $n = 0$ then (i), (ii), (iii) are obvious. Let us assume (i) to (iii) to be true for some n. If $\gamma_n = \tilde{\gamma}_n \neq 0$ then $Y_n = V_1 \cup V_2 = \tilde{Y}_n = \tilde{V}_1 \cup \tilde{V}_2$ where the boxes arising by bisection are equal, $v_j = \tilde{V}_j$ for $j = 1, 2$. Let $v_j = \mathrm{lb}F(V_j)$ and let $\tilde{v}_j = \mathrm{lb}H(\tilde{V}_j)$. After (Y_n, y_n) and $(\tilde{Y}_n, \tilde{y}_n)$ have been discarded from the list, (V_j, v_j) and $(\tilde{V}_j, \tilde{v}_j)$ are inserted at positions that depend on the size of $v_j, \tilde{v}_j$ and on the manner by which Alg. 1 handles the order when some values are equal. If $v_j < f^*$ then $v_j = \tilde{v}_j$ and the pairs (V_j, v_j) and $(\tilde{V}_j, \tilde{v}_j)$ enter the lists at the same position. If $v_j \geq f^*$ then $\tilde{v}_j = f^*$ and the corresponding pairs need not enter the lists at the same position. In any case, $\gamma_{n+1} = \tilde{\gamma}_{n+1}$ where γ_{n+1} can take one of the values $\gamma_n - 1, \gamma_n$, or $\gamma_n + 1$. If $\gamma_n = \tilde{\gamma}_n = 0$ then $\gamma_{n+1} = \tilde{\gamma}_{n+1} = 0$

and $y_{n+1} = \tilde{y}_{n+1} = f^*$ due to the properties of F and H mentioned above. □

We shall show that the convergence of the sequence (y_n) can be arbitrarily slow even though the convergence order of F is arbitrarily high, and even though F is assumed to satisfy

$$w(F(Y)) \leq cw(Y)^\alpha \text{ for any } Y \in \mathbf{I}(X) \qquad (3.15)$$

for some $a, c > 0$ instead of (3.14). Condition (3.15) is much stronger than (3.14). The reason for this behavior is that the sequence $(w(Y_n))$ can converge so slow that a high order of F cannot restore a reasonable convergence order of (y_n).

It is first necessary to develop some technical preliminaries. For example, we provide an appropriate expression so that the sequences (y_n) generated by Alg. 1 will have an arbitrarily slow speed of convergence. Let S be a collection of convergent sequences $(y_n)_{n=0}^\infty$, where $y_n \to y^*$, say. Let S_0 be the corresponding collection of null-sequences $(y_n - y^*)_{n=0}^\infty$ for $(y_n)_{n=0}^\infty$ belonging to S. Let now (x_n) be some given null-sequence (not necessarily belonging to S) and let $\hat{x}_n = \max_{v \geq n} |x_v|$.
We say that S has the *convergence order* of (x_n) iff for any sequence $(y_n) \in S_0$ a $\rho > 0$ exists such that $|y_n| < \rho \hat{x}_n$ for all $n \in \mathbf{N}$. This means that $|y_n| = \mathcal{O}(\hat{x}_n)$. (If $|x_n|$ instead of $\hat{x}_n$ is used in this definition then difficulties could arise if zeros occur in the sequence.)

If there is no null-sequence having the convergence order of S then we say that *S converges arbitrarily slowly*. Since S has the convergence order of (x_n) iff S has the convergence order of $(\hat{x}_n)$ and since $(\hat{x}_n) \searrow 0$ (that means that $(\hat{x}_n)$ is a monotonically decreasing null-sequence), we can restrict ourselves to monotonically decreasing null-sequences (x_n) as test sequences. A few logical and analytical rearrangements which will not be repeated here lead to the following lemma:

LEMMA 5 *S converges arbitrarily slowly iff given any null-sequence $(x_n) \searrow 0$, a sequence $(y_n) \in S$ with a subsequence $(y_{t_k})_{k=0}^\infty$ can be found such that $|y_{t_k}| \geq k x_{t_k}$ holds for any $k \in \mathbf{N}$.* □

The following assertion is also obvious:

LEMMA 6 *If* $(x_n) \searrow 0$ *then a subsequence* $(x_{v_n})_{n=1}^{\infty}$ *exists such that* $(nx_{v_n})_{n=1}^{\infty} \searrow 0$. $\square$

For technical reasons in the proof, we introduce a property (P) that is sufficient for S to be arbitrarily slow: We say that S has property (P) if *for any positive reals α, c and any increasing sequence of positive integers, $(s_k)_{k=1}^{\infty}$, a sequence $(y_n) \in S$ can be found such that, for any positive integer k, an integer $t_k \geq s_k$ exists so that $|y_{t_k}| \geq c2^{-k\alpha}$.*

LEMMA 7 *If (P) holds for S then S converges arbitrarily slowly.*

Proof. The assertion is shown via Lemma 5: Let $(x_n) \searrow 0$ be given. Then, by Lemma 6, a sequence of indices $(v_k)_{k=1}^{\infty}$ exists, so that $(kx_{v_k})_{k=1}^{\infty} \searrow 0$. Let $\alpha, c > 0$ be chosen arbitrarily, and set $\xi_k = c2^{-\alpha k}$. Then $(\xi_k)_{k=1}^{\infty} \searrow 0$. Since (kx_{v_k}) is also a null-sequence it has a subsequence which is majorized by $(\xi_k)_{k=1}^{\infty}$. That is, there exists a subsequence $(m_k)_{k=1}^{\infty}$ of $(v_k)_{k=1}^{\infty}$ so that $\xi_k \geq lx_{m_k}$ and $m_k = v_l$, where l depends on k. That is, we have $m_k \geq v_k$ and $l \geq k$. Set $s_k = m_k$ for any k. Then, by (P), a sequence $(y_n) \in S_0$ exists such that for each $k \geq 1$ some $t_k \geq s_k$ can be found with $|y_{t_k}| \geq \xi_k$. The sequence (y_n) and its subsequence (y_{t_k}) are then the sequences required for applying Lemma 4, since we have

$$\xi_k \geq lx_{m_k} = lx_{s_k} \geq lx_{t_k} \geq kx_{t_k}. \quad \square$$

Let us now turn to the main result of this section, which is that Alg. 1 may approach its solution, that is, the global minimum arbitrarily slowly.

Let $X \in \mathbf{I}^m$ and $\alpha, c > 0$ be given. Let S be the set of all sequences of lower bounds $(y_n)_{n=0}^{\infty}$, which are generated by applying Alg. 1 to functions $f : X \to \mathbf{R}$ and their inclusion functions $F : \mathbf{I}(X) \to \mathbf{I}$ satisfying (3.15). The functions f are therefore continuous, the inclusion functions are of order α and satisfy the condition $w(F(Y)) \to 0$ as $w(Y) \to 0$, and thus all the sequences (y_n) converge by Theorem 3.

THEOREM 5 *If S is the sequence set defined above then S converges arbitrarily slowly.*

Proof. Without loss of generality, we choose $X = [0,1]$. This implies $m = 1$. We apply Alg. 1 to the constant function $f(x) = 0$ for $x \in X$. Thus, $S = S_0$. We will show that S has property (P). Then the theorem follows by Lemma 7.

Let $U_k = [1 - 2^{-k}, 1]$ and $x_k = c^{-k\alpha}$ for positive integers k. Let $(s_k)_{k=1}^{\infty}$ be an increasing sequence of positive integers. An inclusion function satisfying (3.15) is

$$F(Y) = [-c2^{-\alpha(s_k+\frac{1}{2})}, 0] \text{ if } Y = U_k \text{ for some } k > 0$$
$$= [-cw(Y)^{\alpha}, 0], \text{ otherwise.}$$

We will show that S^* satisfies (P) where S^* consists only of the sequence of lower bounds generated by F. Then (P) holds also for S since $S^* \subseteq S$.

We first consider the case $k = 1$, that is $s_k = s_1, t_k = t_1$. We get $Y_0 = X, y_0 = -c$. Then Y_0 is bisected into $Y_1 = [0, 2^{-1}]$ and $U_1 = [2^{-1}, 1]$. Furthermore, we have $y_1 = -c2^{-\alpha}$ and $u_1 := \mathrm{lb}F(U_1) = -c2^{-\alpha(s_1+\frac{1}{2})} > y_1$. Now Alg. 1 induces a uniform subdivision of Y_1 in subboxes $Y \subseteq Y_1$ as long as $y = \mathrm{lb}F(Y) > u_1$. The last box Y with $y < u_1$ has width $w(Y) = 2^{-s_1}$, and thus $y = -c2^{-\alpha s_1}$. This box shall be indexed by $t_1 - 2$. That is, t_1 can be explicitly counted or determined. The leading box of the subsequent list is $Y_{t_1-1} = U_1 = [2^{-1}, 1]$ with $y_{t_1-1} = u_1 = -c2^{-\alpha(s_1+\frac{1}{2})}$. Then U_1 is bisected into $U_{11} = [\frac{1}{2}, \frac{3}{4}]$ and $U_{12} = U_2$ with $u_{11} = \mathrm{lb}U_{11} = -c2^{-2\alpha}$ and $u_{12} = \mathrm{lb}U_{12} = -c2^{-\alpha(s_2+\frac{1}{2})}$. Thus, $Y_{t_1} = U_{11}$ and $|y_{t_1}| > x_{t_1}$. Since $t_1 - 2$ is the number of all intervals having the right endpoints smaller than or equal to $\frac{1}{2}$ and which have been generated from X by applying at most s_1 bisections (X included), we have $t_1 > s_1$.

Let us now assume that we have already determined $t_{k-1} \geq s_{k-1}$ where $k \geq 2$. In order to get $t_k \geq s_k$ we choose the index t_k so that $Y_{t_k-1} = U_k$. This means that Y_{t_k-2} is the last leading box which arises from X using s_k bisections. Therefore $y_{t_k-1} = -c2^{-\alpha(s_k+\frac{1}{2})}$, and Y_{t_k-1} is bisected into

$$U_{k1} = Y_{t_k} = [1 - 2^{-k}, 1 - 2^{-k-1}], \ U_{k2} = [1 - 2^{-k-1}, 1],$$

where $|y_{t_k}| = c2^{-\alpha(k+1)} > c2^{-\alpha k} = x_k$. Since $t_k - 2$ is larger than the number of all boxes Y which satisfy $\mathrm{ub}\, Y \leq \sum_{v=1}^{k} 2^{-v}$ and arise from X using at most s_k bisections we get $t_k > s_k$. □

3.7 Convergence Speed with Isotone Inclusion Functions

In Sec. 3.6 we have seen that the Moore-Skelboe algorithm can be arbitrarily bad if the inclusion functions are only restricted by order conditions. In this section it is shown that the use of isotone inclusion functions leads to reasonable convergence results. First some technical properties are derived that are needed for the main results. That is, in order to make assertions about the convergence order of Alg. 1 we have to study the worst possible case as mentioned in Sec. 3.6.

We begin by showing that isotone inclusion functions causing a uniform subdivision of X give the slowest possible approach to f^*. For that reason, two inclusion functions F and G for f are compared. We assume that F and G have been transformed from isotone inclusion functions of order α by the simplification process described in Sec. 3.6. According to Lemma 3, the constant function $f(x) = 0$ for all $x \in X$ can be considered. Thus, $f^* = 0$ and $w(\Box f(Y)) = 0$ for all $Y \in \mathbf{I}(X)$. We set

$$F(Y) = [-cw(Y)^{\alpha}, 0] \text{ for } Y \in \mathbf{I}(X)$$

whereas $G : \mathbf{I}(X) \to \mathbf{I}$ is assumed to satisfy

$$w(G(Y)) \leq cw(Y)^{\alpha} \text{ for } Y \in \mathbf{I}(X).$$

The algorithm applied to F and G generates lists that shall be denoted by

$$\begin{aligned} \mathbf{L}_n &= ((Z_{ni}, z_{ni}))_{i=1}^{n+1} \quad \text{for } F, \\ \overline{\mathbf{L}}_n &= ((\overline{Z}_{ni}, \overline{z}_{ni}))_{i=1}^{n+1} \quad \text{for } G. \end{aligned}$$

We set $Y_n = z_{n1}, y_n = \mathrm{lb} F(Y_n), \overline{Y}_n = \overline{Z}_{n1}, \overline{y}_n = \mathrm{lb} G(\overline{Y}_n)$. We have

$$(y_n) \nearrow f^* \text{ and } (\overline{y}_n) \nearrow f^* \tag{3.16}$$

which follows from Lemma 4 or, in case of F, from the isotonicity of F directly. The notation $(y_n) \nearrow f^*$ shall mean (y_n) is a monotonically increasing sequence converging to f^*.

Considering F and G, it seems self-evident that F leads to a worse convergence approach to f^* than G. A proof is, however, necessary since it cannot be excluded in advance that for some n or several n or infinitely many n,

$$\overline{y}_n < y_n,$$

which could arise if Y_n is small and $\overline{Y}_n$ is large.

LEMMA 8 *Let F, G and the lists $L_n, \overline{L}_n$ be as defined above. Then $y_n \leq \overline{y}_n$ for all n.*

Proof. It is assumed that $c = \alpha = 1$ without restricting the generality. We use mathematical induction with respect to the list indices n. The case $n = 0$ is obvious. Let us now assume that

$$y_k \leq \overline{y}_k \text{ for all } k \leq n \tag{3.17}$$

holds. We have to show that $y_{n+1} \leq \overline{y}_{n+1}$. Two cases are to be distinguished:

(i) $w(Y_n) = w(Y_{n+1})$,

(ii) $w(Y_{n+1}) < w(Y_n)$.

Case (i) implies $y_{n+1} = y_n$. Again by (i) and by (3.16) we get

$$y_{n+1} = y_n \leq \overline{y}_n \leq \overline{y}_{n+1},$$

which proves the assertion in case (i). For dealing with (ii) let $(w_v)_{v=0}^{\infty}$ denote the *decreasing ordered* sequence of *different* box widths that occur in the bisection process of X. Thus, for example, $w_0 = w(X)$, $w_1 = w(X)/2$, etc. At any bisection of a box $Y = V_1 \cup V_2$ with $w(Y) = w_v$, we find that $w(V_1) = w(V_2)$ which is equal to w_v or w_{v+1}. If $w(V_1) = w_{v+1}$ we call V_1 a *fresh* box. Conversely, if V arises by bisection from Y, and $w(V) = w_v$, then $w(Y) = w_{v-1}$ if V is fresh, $w(Y) = w_v$ otherwise. (ii) is now equivalent to the existence of a v such that only boxes of widths w_v and w_{v+1} occur in L_n, and all boxes of L_{n+1} are of width w_{v+1} and fresh. That is,

$$\left.\begin{array}{ll} L_n: & y_n = -w_v < z_{n2} = \ldots = z_{n,n+1} = -w_{v+1}, \\ L_{n+1}: & y_{n+1} = z_{n+1,2} = \ldots = z_{n+1,n+2} = -w_{v+1}. \end{array}\right\} \quad (3.18)$$

We will show that the assumption

$$y_{n+1} > \overline{y}_{n+1} \quad (3.19)$$

leads to a contradiction. (3.19) implies $w(\overline{Y}_{n+1}) > w(Y_{n+1}) = w_{v+1}$. According to the properties of the widths-sequence, it follows that

$$w(\overline{Y}_{n+1}) \geq w_v.$$

Thus, $\overline{Y}_{n+1}$ was obtained from X by fewer bisections than each box of L_{n+1}; see (3.19). Since L_{n+1} and $\overline{L}_{n+1}$ are obtained by the same number of bisections, at least one box of $\overline{L}_{n+1}$, say $\overline{V}$, is obtained from X by more bisections than each box of L_{n+1}. Since each box of L_{n+1} has suffered from the same number of bisections, we have either $w(\overline{V}) = w_v$ or $w(\overline{V}) = w_{v+1}$, but in this second case, $\overline{V}$ is not fresh. If $\overline{V}$ is obtained by (one) bisection of some box $\overline{U}$ of some former list, we have

$$w(\overline{U}) \leq w_{v+1}.$$

If $\overline{u} = \text{lb}G(\overline{U})$, then

$$\overline{u} \geq -w_{v+1} = y_{n+1}.$$

Further, because of (3.19), we have

$$\overline{y}_{n+1} < y_{n+1} \leq \overline{u}$$

and finally, by (3.16),

$$\overline{y}_k < \overline{u} \text{ for all } k \leq n+1.$$

This implies that $\overline{U}$ has never been a leading box, which would have been necessary to get $\overline{U}$ bisected as was assumed. This provides the contradiction. □

In order to further narrow down the worst possible convergence case we show that among all boxes of constant width, cubes yield the slowest convergence if Alg. 1 is applied and the worst isotone inclusion function - see the previous lemma - is used. For this purpose, let $X \in \mathbf{I}^m, w(X) = 1$ and $X = [0,1]^m$. Let further $G : \mathbf{I}(\overline{X}) \to \mathbf{I}$ and $F : \mathbf{I}(X) \to \mathbf{I}$ be defined formally by the same expression

$$G(Y) = F(Y) = [-cw(Y)^\alpha, 0].$$

Thus, F and G are both isotone inclusion functions for the constant function $f(x) = 0$. The lists relating to F and G shall be denoted by $L_n = ((Z_{ni}, z_{ni}))_{i=1}^{n+1}$ and $\overline{L}_n = ((\overline{Z}_{ni}, \overline{z}_{ni}))_{i=1}^{n+1}$, respectively, and we set $Y_n := Z_{n1}, y_n := z_{n1}, \overline{Y}_n := \overline{Z}_{n1}$, and $\overline{y}_n := \overline{z}_{n1}$. It is necessary to make Alg. 1 more precise. The order of the lists is regulated by Step 10 of the algorithm. If Step 10 cannot decide uniquely on the ordering then the box of larger width shall precede the box of smaller width. If this does not result in a decision then that box shall precede which has the larger edge-widths sum.

LEMMA 9 *Under the above conditions $y_n \leq \overline{y}_n$ holds for all $n \in \mathbf{N}$.*

Proof. It is shown that for any $n \in \mathbf{N}$ an $i \in \mathbf{N}$ exists such that the relations

$$\overline{Y}_n = \overline{Z}_{n2} = \ldots = \overline{Z}_{ni} \supseteq \overline{Z}_{n,i+1} = \ldots = \overline{Z}_{n,n+1},$$
$$Y_n = Z_{n2} = \ldots = Z_{ni} \supseteq Z_{n,i+1} = \ldots = Z_{n,n+1},$$
$$\overline{Y}_n \subseteq Y_n, \qquad \overline{Z}_{n,i+1} \subseteq Z_{n,i+1}$$

hold. $A = B$ or $A \subseteq B$ mean equality or inclusion after an appropriate motion of A or B. If mathematical induction is applied then the relations are obvious. Then, Lemma 9 follows immediately. □

We can now give sharp upper bounds for the convergence order of Alg. 1 under the assumption that the inclusion functions are isotone.

Let again $X \in \mathbf{I}^m, f : X \to \mathbf{R}$ and let $F : \mathbf{I}(X) \to \mathbf{I}$ be an inclusion function for f. Furthermore let $\omega = w(X)$. Let (Y_n, y_n) denote the leading pairs if the algorithm is applied to F and X.

The following theorem shows that Alg. 1 renders convergence to f^* of the order of the sequence $(n^{-\alpha/m})_{n=0}^\infty$ if α is the order of the inclusion function. One may notice that the slowness of the algorithm caused when the dimension m is higher can be compensated – at least theoretically – by using a higher order inclusion function.

THEOREM 6 *Let the inclusion function F for f be isotone and satisfy $w[F(Y)] - w[\Box f(Y)] \leq cw(Y)^\alpha$ for all $Y \in \mathbf{I}(X)$ and some constants $c, \alpha > 0$. Then*

$$f^* - y_n \leq c(2w)^\alpha (n+2)^{-\alpha/m},$$

that is, $f^* - y_n = \mathcal{O}(n^{-\alpha/m})$.

Proof. Referring to the preceding lemmas we can assume that $f(x) = 0$ for all $x \in X$, that $F(Y) = [-cw(Y)^\alpha, 0]$, and that X is a cube of width w. These assumptions indicate the worst possible case. Since X is a cube, the maximum box width of the boxes of the lists is w and therefore $y_n = -cw^\alpha$ for $n = 0, \ldots, 2^m - 2$. The maximum box width is $w/2$ and $y_n = -cw^\alpha/2^\alpha$ for $n = 2^m - 1, \ldots, 2^{2m} - 2$. In general, the maximum box width is $w/2^l$ and $y_n = -cw^\alpha/2^{l\alpha}$ for the lists L_n where $n = 2^{lm} - 1, \ldots, 2^{(l+1)m} - 2$ and $l \in \mathbf{N}$.

F is isotone. The sequence (y_n) is therefore increasing. This gives

$$\begin{aligned} y_n &\geq -cw^\alpha/2^{l\alpha} \text{ if } n \geq 2^{lm} - 1, \\ y_n &\leq -cw^\alpha/2^{l\alpha} \text{ if } n \leq 2^{(l+1)m} - 2. \end{aligned}$$

Some rearrangements show that

$$\begin{aligned} n &\geq 2^{lm} - 1 \text{ iff } -c(n+1)^{-\alpha/m} \geq -c/2^{l\alpha}, \\ n &\leq 2^{(l+1)m} - 2 \text{ iff } -c(n+2)^{-\alpha/m} 2^\alpha \leq -c/2^{l\alpha}. \end{aligned}$$

Therefore, $n \in [2^{lm} - 1, 2^{(l+1)m} - 2]$ implies

$$-c(2w)^\alpha (n+2)^{-\alpha/m} \leq y_n \leq -cw^\alpha (n+1)^{-\alpha/m}. \tag{3.20}$$

This implication holds for any $l \in \mathbf{N}$. Conversely, if $n \in \mathbf{N}$ is given then there exists exactly one l such that both parts of the implication are satisfied. It is defined by $y_n = -cw^\alpha/2^{l\alpha}$. Using this formula, the parameter l can be eliminated, and (3.20) holds in general. Thus, constants d and $\tilde{d}$ exist such that

$$-dn^{-\alpha/m} \leq y_n = y_n - f^* \leq -\tilde{d}n^{-\alpha/m}.$$

That is, $f^* - y_n = \mathcal{O}(n^{-\alpha/m})$. $\Box$

Remark. There exist infinitely many numbers n such that the left-hand inequality of (3.20) is an equality. In this sense, the order estimate of Theorem 6 is sharp.

Example *for the fastest approach not using acceleration devices.* Theorem 6 gives an upper bound for the convergence speed of the y_n to f^*. Which is the fastest approach that can be expected using Alg. 1? The fastest speed is obviously given if only *one* global minimizer exists and if this minimizer is contained in all leading boxes. Let, for simplicity, X again be a cube, and $w(X) = w$. (A more general shape of X would only change the multiplicative constant occurring in the order estimate slightly, but not the exponential term.) In order to eliminate the direct influence of the special shape of f or F, we only assume $w(F(Y)) \leq cw(Y)^\alpha$ for $Y \in \mathbf{I}(X)$ and for some constants $c, \alpha > 0$. In contrast to the rather high number of bisections shown in the proof of Theorem 6, we have that

$$lm \leq n \leq (l+1)m - 1 \text{ implies } w(Y_n) = w/2^l (l \in \mathbf{N}).$$

If $[q]$ denotes the largest integer smaller than or equal to q, we get

$$w(Y_n) = w/2^{[n/m]}.$$

If $n/m = [n/m] + r_n$ then $r_n \in [0, 1 - 1/m]$. Finally,

$$\begin{aligned} f^* - y_n &\leq w(F(Y_n)) - w(\Box f(Y_n)) \leq cw(Y_n)^\alpha \\ &= cw^\alpha / 2^{\alpha[n/m]} \leq c(2^{r_n} w)^\alpha (2^{\alpha/m})^{-n}. \end{aligned}$$

This leads to the following.

Remark. In general, the fastest convergence speed of (y_n) to f^* under the conditions specified above can be estimated by

$$f^* - y_n \leq c(2^{r_n} w)^\alpha (2^{\alpha/m})^{-n},$$

that is,

$$f^* - y_n = \mathcal{O}((2^{\alpha/m})^{-n}).$$

Example. If $m = 2, \alpha = 2$ (2 variables, F is of order 2) then one can expect the sequence $f^* - y_n$ to be between $d/2^n$ and $\tilde{d}/n$ for some constants $d, \tilde{d}$. If $m = 1, \alpha = 2$ (1 variable, F is of order 2) then one can expect the sequence $f^* - y_n$ to be between $d/4^n$ and $\tilde{d}/n^2$ for some constants $d, \tilde{d}$.

3.8 Ichida-Fujii Algorithm and its Convergence Conditions

It was shown in the previous sections that Alg. 1 is a simple method for approaching some global minimum points. There are two disadvantages: First, there does not exist an error estimate for this approach, and second, it cannot be guaranteed that all minimum points can be reached.

In this section, Alg. 2 (Ichida-Fujii (1979)) will be discussed. It is a modification of Alg. 1 which makes use of a midpoint test and where all the boxes on the list are employed to include the minimum points. The convergence properties of this algorithm have been investigated by Moore-Ratschek (1987). An error estimate for the convergence to the minimum points via the solution set of this algorithm can be given.

Further, no minimum points are lost as is the case with Alg. 1. However, it is theoretically possible that a proper superset of X^*, the set of global minimizers, is gained. This possibility is, in practice, extremely unlikely.

ALGORITHM 2 *arises from Algorithm 1 by inserting the following step* 11^+ *between steps 11 and 12 of Alg. 1:*

11^+. *Discard all pairs (Z, z) from the list that satisfy $F(c) < z$ where $c = \text{mid } Y$.*

Step 11^+ is called a *midpoint test* and is used to reduce the number of boxes of the lists $\mathbf{L}_n$. Obviously, the test remains valid if $F(c) < z$ is replaced by $F(c') < z$ for any $c' \in Y$ or by $\text{ub} F(Y) < z$. The updating of the current values y_n and f_n is here again recommended, cf. Sec. 3.3.

We again assume that the termination criteria provided in the algorithm are missing or will never cause the algorithm to stop, so that the convergence properties of Alg. 2 can be studied.

We continue to consider a box $X \in \mathbf{I}^m$, a function $f : X \to \mathbf{R}$ and an inclusion function $F : \mathbf{I}(X) \to \mathbf{I}$ for f. Let X^* be the set of global minimizers of f in X and U_n the union of all boxes that occur in the n-th list L_n generated by Alg. 2, i.e.,

$$U_n = \bigcup_{i=1}^{l_n} Z_{ni}$$

where l_n denotes the length (number of pairs) of the list L_n. Clearly we always have $l_n \leq n+1$. The unions U_n are compact sets. The following theorem provides the main property of Alg. 2, that is, the unions U_n converge monotonically to a superset, B, of X^* from the exterior. The convergence is based upon the distance d introduced in Sec. 3.2. Since Alg. 2 has all the features of Alg. 1, the properties of Alg. 1 mentioned in the last section are also valid for Alg. 2. Since under very natural conditions the y_n converge to f^*, we obtain the solution set of Alg. 2 in B and f^*. In contrast to Alg. 1 where the only use of the midpoint test was to reduce the list, the test is now advanced to be a basic step of the algorithm, that is, it has an influence on the solution set. The reason is that Alg. 2 considers all boxes of the list L_n and Alg. 1 only the leading boxes Y_n which are never touched by the midpoint test.

THEOREM 7 *Let the inclusion function F of f satisfy $w(F(Y)) \to 0$ as $w(Y) \to 0$. Then there exists a set $B \supseteq X^*$ such that $U_n \supseteq B$ for all n and $U_n \to B$ as $n \to \infty$. The sequence (U_n) is nested such that*

$$B = \bigcap_{n=1}^{\infty} U_n.$$

Proof. The assertion of the theorem is satisfied by $B = \bigcap_{n=1}^{\infty} U_n$. We first show that $X^* \subseteq U_n$ for any n so that $X^* \subseteq B$ follows. In Alg. 1 one box was bisected in each iteration and then it was replaced by its parts. Therefore we had, for any n,

$$U_n = \bigcup_{i=1}^{n} Z_{ni} = X.$$

After having counted the lists temporarily from 0 in Sec. 3.6 and 3.7 we now revert to counting the lists from $n = 1$. We now consider Alg. 2. The discarding process which was incorporated, Step 11^+, removes a box Z_{ni} when $F(c_n) = f(c_n) < z_{ni} = \min F(Z_{ni})$ where c_n is the midpoint of the leading box Y_n. Since $f^* < f(c_n)$ and $\Box f(Z_{ni}) \subseteq F(Z_{ni})$, the box Z_{ni} cannot contain a global minimizer. Thus all boxes of any list that contain a global minimum point remain in the list, and $X^* \subseteq U_n$ for any n.

Second, the convergence of the unions U_n to B follows directly from the definition of B and the chain property of these unions, that is, $U_1 \supseteq U_2 \supseteq \ldots$. $\Box$

How large is the overestimation of X^* by B? The discarding process seems to be so strong that one would not expect other points than global minimum points to remain in X^*. This view is further strengthened by the examples of Ichida-Fujii (1979). There is, however, one exception which may arise, at least theoretically, but this fact prevents writing X^* instead of B in Theorem 7. This exception can occur if there exists some *nonleading* box z_{ni} such that

$$f^* = z_{ni} = \min F(Z_{ni}).$$

Thus, z_{ni} will almost never be a leading box and has no chance of getting bisected and producing values strictly larger than f^*. Further, the inequality

$$F(c_n) < z_{ni}$$

will never hold, and so z_{ni} will never be discarded by Step 11^+. Therefore, z_{ni} is contained in all the lists. This implies $Z_{ni} \subseteq B$.

The choice of termination criteria, error estimations, and the influence of rounding errors is dealt with in Sec. 3.10.

3.9 Hansen's Algorithm and its Convergence Conditions

The following modification of Alg. 2 for solving the unconstrained problem (3.1), due to Hansen, makes it possible to give a *mathematically* satisfactory answer to the uncertainty inherent in Algorithms

1 and 2. Hansen does not use an ordering of the list with respect to the $z_{ni} = \text{lb}F(Z_{ni})$ as is done in Alg. 1 and Alg 2. He orders the list either with respect to the widths of the boxes Z_{ni} or to the age of these boxes when the counting of the age begins at the generation of a box by a bisection. Both versions lead to a uniform subdivision of all boxes which have not been discarded yet. Error estimation via the Lebesgue measure of the solution is again possible; see Sec. 3.10.

The input parameters for the algorithm are the box X, the inclusion function F of f, and some accuracy parameter etc., which may be needed for the termination criteria.

ALGORITHM 3 *(Hansen)*

1. *Set $Y := X$.*
2. *Calculate $F(Y)$ and $\tilde{f} := \text{ub}F(c)$ where $c = \text{mid } Y$*
3. *Set $y := \text{lb}F(Y)$.*
4. *Initialize list $\mathbf{L} := \{(Y, y)\}$.*
5. *Choose a coordinate direction k parallel to which Y has an edge of maximum length, that is, $k \in \{i : w(Y_i) = w(Y)\}$.*
6. *Bisect Y normal to direction k getting boxes V_1, V_2 such that $Y = V_1 \cup V_2$.*
7. *Calculate $F(V_1), F(V_2)$.*
8. *Set $v_i := \text{lb}F(V_i)$ for $i = 1, 2$.*
9. *Enter the pairs $(V_1, v_1), (V_2, v_2)$ at the end of the list.*
10. *Choose a pair $(\tilde{Y}, \tilde{y})$ of the list which satisfies $\tilde{y} \leq z$ for all pairs (Z, z) of the list.*
11. *Discard all pairs (Z, z) from the list that satisfy $\tilde{f} < z$ (midpoint test).*
12. *If termination criteria hold go to 15.*
13. *Denote the first pair of the list by (Y, y). Set $c := \text{mid } Y$ and $\tilde{f} := \min(\tilde{f}, \text{ub}F(c))$.*
14. *Go to 5.*

15. End.

As was the case in Algorithm 2, the inequalities $F(c) < z$ which are involved in the midpoint test can be replaced by $F(c') < z$ for any $c' \in Y$. The updating of $\tilde{y}$ is again recommended, cf. Sec. 3.3. An updating of $\tilde{f}$ is incorporated in the algorithm.

The algorithm uses an ordering of the lists with respect to the age of their boxes. The ordering with respect to the widths of the boxes leads to an algorithm, denoted by *Algorithm 3+*, which arises from Alg. 3 by replacing Step 9 by

Step 9+. Enter the pairs (V_1, v_1) and (V_2, v_2) into the list such that the widths of the boxes in the list decrease (not necessarily strictly).

Alg. 3 is certainly a special case of Alg. 3+ since Step 9+ leaves the ordering open if boxes have the same widths. When however acceleration devices are incorporated, Alg. 3 and 3+ can be essentially different since these devices can diminish the width of a box.

The crucial point of these two algorithms is that all boxes occurring on the lists will be bisected periodically or discarded.

The pairs $(\tilde{Y}, \tilde{y})$, cf. Step 10, occupy a key position even when they seem to be without influence on the algorithm. These pairs were the leading pairs of Alg. 1 and Alg. 2 and they are the only boxes, Z, of the lists for which $f^* \in F(Z)$ is guaranteed. If $\tilde{y}_n$ is the value that $\tilde{y}$ takes at the n-th iteration then the sequence $(\tilde{y}_n)$ plays the same role in Alg. 3 that the sequence (y_n) plays in Alg. 1 and 2. This means that, under the condition (3.12), we get $\tilde{y}_n \to f^*$ and $\tilde{y}_n \leq f^*$.

Let Y_n be the leading box and Z_{ni} an arbitrary box of L_n which is the list generated at the n-th iteration of Alg. 3. Not only does the contraction property

$$w(Y_n) \to 0 \text{ as } n \to \infty$$

hold in case of Alg. 3, but even

$$w(Z_{ni}) \to 0 \text{ as } n \to \infty$$

is satisfied. This follows from the fact that the lists are ordered with respect to the ages or the widths of the boxes.

We keep in mind that (3.12), that is the condition $w(F(Y)) \to 0$ as $w(Y) \to 0$, implies the continuity of f but not a Lipschitz condition of f.

The following theorem (due to Moore-Ratschek (1987)) states that the unions of the boxes of the lists contract to exactly the set of global minimum points, X^* of f on X. So X^* and f^* is the solution set of Alg. 3. The notation and assumptions are the same as in Sec. 3.8. The union of the boxes of the list L_n which is generated by Alg. 3 is again denoted by U_n.

THEOREM 8 *Let the inclusion function F of f satisfy $w(F(Y)) \to 0$ as $w(Y) \to 0$. Then $U_n \supseteq X^*$ for all n and $U_n \to X^*$ as $n \to \infty$. The sequence (U_n) is nested and thus $X^* = \bigcap_{n=1}^{\infty} U_n$.*

Proof. Since the discarding process is the same as in Theorem 7, the result (see the proof of Theorem 7)

$$X^* \subseteq \bigcap_{n=1}^{\infty} U_n$$

holds. This inclusion is not affected by the fact that the order of the lists is not the same as in Theorem 7. The inclusion $\bigcap_{n=1}^{\infty} U_n \subseteq X^*$ in the opposite direction remains to be proved: Let w_n denote the maximum width of the boxes of the n-th list, L_n. Then

$$w_n \to 0 \text{ as } n \to \infty. \tag{3.21}$$

Now, let us assume that $x \in U_n$ for all n. We have to show that $x \in X^*$. It first follows that, for any n, a pair (Z'_n, z'_n) occurs in L_n such that $x \in Z'_n$. Since this pair has not yet been discarded, the condition

$$f(c_n) \geq z'_n \tag{3.22}$$

holds owing to the midpoint test where $c_n = \text{mid } \tilde{Y}_n$ and $(\tilde{Y}_n, \tilde{y}_n)$ is a pair with minimal $\tilde{y}_n$. The sequence (Z'_n) tends to x, since $x \in Z'_n$ for all n and $w(Z'_n) \to 0$ as $n \to \infty$, due to (3.21). Therefore $w(F(Z'_n)) \to 0$, from the assumptions of the theorem. Furthermore,

$\Box f(Z'_n) \to f(x)$, since f is continuous. Applying the derived convergence properties to.

$$f(x) \in \Box f(Z'_n) \subseteq F(Z'_n),$$

we get

$$F(Z'_n) \to f(x) \text{ as } n \to \infty. \qquad (3.23)$$

Finally we prove

$$F(Z'_n) \to f^* \text{ as } n \to \infty;$$

this implies, together with (3.23), that $f^* = f(x)$ and $x \in X^*$.

We consider the inequalities

$$\tilde{y}_n \leq z'_n \leq f(c_n) \in F(\tilde{Y}_n).$$

The first inequality is due to the definition of the pair $(\tilde{Y}_n, \tilde{y}_n)$; the second is (3.22). Since $F(\tilde{Y}_n) \to f^*$ and $\tilde{y}_n \to f^*$ (see Theorem 2), we have $z'_n \to f^*$, which implies, together with $w(F(Z'_n)) \to 0$, the assertion, that is, $F(Z'_n) \to f^*$. $\Box$

Theorem 8 could also be derived from Basso's (1982) Theorem 1. In this case one had first to prove that his assumption (H) is satisfied by Algorithm 3. In fact, condition (3.23) implies (H).

As we see, the convergence proofs of Theorem 8 and of previous theorems do not depend on the number of global minimum points of f. This number may be finite or infinite, and in the second case, the minimum points may occur isolated or as continua. Each global minimum point is contained in at least one box and in at most $2^m \cdot s$ boxes of any list. In general, a point is contained in several boxes if it is an edge point of some box, and it is contained in $2^m \cdot s$ boxes if it is a corner of some box. This implies the following statement.

COROLLARY 1 *If f has finitely many global minimizers, say s, then the number of boxes of the lists $\mathbf{L}_n$ tends to some number $k \leq 2^m \cdot s$ as $n \to \infty$. Otherwise, the number tends to ∞.* $\Box$

In the case of s global minimum points the probability that a global minimum point lies on the edge of a box is zero and we therefore have $k = s$, in general. The following version of Corollary 1 avoids such distinctions by taking the number of connected components of U_n instead of the number of boxes of the lists. A connected component by U_n is the union of a maximum number of boxes Z_{ni} of the list L_n such that each two of these boxes can be connected by a (continuous) path which lies totally in U_n.

COROLLARY 2 *If f has $s < \infty$ global minimum points and if γ_n denotes the number of connected components of U_n, then $\gamma_n \to s$ as $n \to \infty$.* □

The meaning of the corollaries for *practical* calculations should not be overestimated. The number s (or whether infinitely many minimum points are present) will not be known during a computation, since each computation is terminated after finitely many steps and since the number of minimum points contained in each box of the list when the calculation is terminated is unknown. Further, the number of boxes of a list as well as the number of connected components can be smaller and also be greater than s at any time of the computation. On the other hand, the error will not be too large if the number of boxes is identified with s provided the function f is "reasonable" and provided the boxes are sufficiently small.

Therefore it is only correct to make assertions about the number of global minimum points if additional information about f is used. See, for example, the discussion of Shubert's function,

$$f(x) = -\sum_{k=1}^{5} k\sin(k + (k+1)x), \; x \in X = [-10, 10]$$

in Hansen (1979). Hansen's computation results in three boxes,

$$\begin{aligned} Z_1 &= [-6.7745\ 76144, \quad -6.7745\ 76143], \\ Z_2 &= [5.7917\ 89015, \quad 5.7917\ 99064], \\ Z_3 &= [-0.49139\ 21876, \quad -0.49138\ 95811]. \end{aligned}$$

It is tacitly assumed that each of these boxes contains exactly one global minimizer; however, this is not guaranteed. One may confirm this assumption in the following manner using periodicity and convexity arguments as the additional information mentioned: First, according to the mathematical conditions of the problem it is obvious that at least one global minimizer exists which must be located in one of the three boxes. Second, f is of periodicity 2π. So, if $x* \in X^*$ and $x^* \pm 2\pi v \in X$, then $x^* \pm 2\pi v \in X^*, v = 1, 2, \ldots$. Since the distances of the boxes are about 2π, each box contains *at least* one global minimizer. Third, the widths of the boxes are so small that it is almost certain that each box contains just one global minimum point. Nevertheless, in order to guarantee this, the following convexity test was applied: We chose

$$F'''(Y) = \sum_{k=1}^{5} k(k+1)^2 \text{ISIN}(k + (k+1)y)$$

as an inclusion function for f''', where ISIN was an inclusion function for sin, and obtained

$$F'''(Z_3) = [309.91, 309.92].$$

Therefore $f'''(x) > 0$ for all $x \in Z_3$ which means that f is strictly convex in Z_3 and that Z_3 cannot contain 2 global minimum points. The same holds for Z_1 and Z_2 because of the periodicity.

3.10 Termination Criteria, Approximation Errors and Influence of Rounding Errors

The termination of the Moore-Skelboe algorithm via the *convergence to f^**, related error estimates, and problems caused by rounding errors were extensively discussed in Sec. 3.3. These considerations are valid also for the other algorithms of this chapter, i.e. for the leading pairs of the Ichida-Fujii algorithm, and for the pairs $(\tilde{Y}_n, \tilde{y}_n)$ of the Hansen algorithm.

A termination of these algorithms via the *convergence to global minimum points* only using the leading boxes is not appropriate because of the properties discussed in Sec. 3.4. These properties also do not allow an error estimation of the approximation of the global minimizers by the leading boxes.

Therefore we will only be concerned with Alg. 2 and 3 in this section. For a unified treatment we take $B = \bigcap_{n=1}^{\infty} U_n \supseteq X^*$ as the solution set of both algorithms where $B = X^*$ in case of Alg. 3; see Sec. 3.8 and 3.9. The following two termination criteria were provided by Hansen (1979), (1980). If B is at most denumerable and if B is to be included in a set with prescribed accuracy, then

$$\sum_{i=1}^{l_n} w(Z_{ni}) < \epsilon \qquad (3.24)$$

or

$$w(Z_{ni}) < \epsilon \text{ for } i = 1, \ldots, l_n \qquad (3.25)$$

will do, where l_n is the length of L_n. If B is nondenumerable then (3.24) will fail if the Lebesgue measure of B is at least equal to ϵ. Condition (3.25) works independently of the measure of B.

It is typical of the global optimization problem that only poor error estimates for the approach to single global minimizers are available. That is if one picks out any $x \in U_n$ as an approximation for any (unknown) $x^* \in X^*$ then

$$\| x - x^* \|_\infty \leq \max_{y \in U_n} \| x - y \|_\infty$$

is the best estimate which can be derived. The error $\| x - x^* \|_\infty$ can be arbitrarily large in spite of the validity of (3.24) or (3.25). For example, let $Z_{n1} = [1, 1 + \delta]$ and $Z_{n2} = [10, 10 + \delta]$ with $l_n = 2$ for some small $\delta < \epsilon$. Then $\| x - x^* \|_\infty = | x - x^* | \leq 9 + \delta$, where equality holds if $x = 1$ and $x^* = 10 + \delta$, etc. If, however, it is known for some reason that each box Z_{ni} of the final list L_n contains some $x^* \in G$, then the approximation error is

$$\| x - x^* \|_\infty \leq \max_{i=1,\ldots,l_n} w(Z_{ni}). \qquad (3.26)$$

Returning to Shubert's function (see the end of the last section) it is certain that each box contains a global minimizer.

The influence of rounding errors on the computation of U_n does not cause any problems if interval software as described in Ch. 2 is used. In this case the discarding condition (midpoint test) is executed by the computer as

$$\text{ub}\tilde{F}(c_n) < \tilde{z}_{ni} \qquad (3.27)$$

in the non-updated case of Alg. 2 where $\tilde{F}(c_n) \supseteq F(c_n)$ with $f(c_n) \in F(c_n)$ is a numerical approximation of $f(c_n)$ and $\tilde{z}_{ni} = \text{lb}\tilde{F}(Z_{ni})$ where $\tilde{F}(Z_{ni}) \supseteq F(Z_{ni})$ is the numerical approximation of $F(Z_{ni})$. In the updated case, as in Alg. 3, the discarding condition is

$$\min_{i=1,\ldots,n} \text{ub}\tilde{F}(c_n) < \tilde{z}_{ni}. \tag{3.28}$$

Since (3.27) or (3.28) implies the exact discarding condition in Alg. 2 or Alg. 3, no logical flaws arise, so that the set of global minimizers, X^*, is still contained in the union of boxes of any one of the lists. Although rounding errors can prevent the U_n from converging to B, this is not of great practical concern since even with exact arithmetic B is only obtained, in general, after infinitely many iterations of Alg. 2 or 3. This is another reason that Alg. 2 should not be banned, since it also computes, at any iteration, an inclusion of X^*. By the way, Eldon Hansen in person told the first author at a meeting in Columbus, Ohio in September 1987 that for some time he also had been using the ordering with respect to the z_{ni}.

3.11 Accelerating Devices: An Overview

The 3 algorithms we have treated so far are based on the *exhaustion principles*, that is, the principle of removing areas (subboxes of X) which cannot contain a global minimizer. In the same manner we realize that the *branch and bound principle* forms the overlying structure, that is, the areas are processed which have the largest chance to contain a global minimizer. This is a time-consuming process as was evident from the results in Sec. 3.7. It is therefore very important to combine the principles mentioned with techniques for speeding up the computation. In this section we deal with only a few of these techniques in order to demonstrate how they may be combined with the 3 algorithms. The most perfect and sophisticated method is probably the one of Hansen (1979), (1980) which also has been extensively tested (Walster-Hansen-Sengupta (1985)) and which will also be discussed in the forthcoming book by Hansen.

When considering acceleration devices we maintain our modular system as promised in the Preface: that is, we let the user decide which improvements and devices he wishes to include in his particular version of the algorithms.

We mainly recommend the following acceleration devices:

1. Monotonicity test. It is a global technique and discovers boxes Y where f is strictly monotone. The interior of such a box Y - and in general, the box itself - cannot contain a global minimizer such that Y can be removed.

2. Finding a function value as small as possible. The smaller the smallest known or computed function value is at the n-th iteration the more effective is the midpoint test, that is, boxes are removed earlier before they would otherwise have been processed.

For example, let f_n be the smallest function value known up to the n-th iteration. Let $\tilde{f}_n < f_n$ be a smaller value found by an additional procedure. The gain from of this additional effort is that, by the midpoint test, not only the boxes Z_{ni} with

$$f_n < \text{lb}F(Z_{ni})$$

can be discarded (cf. Alg. 2 and 3), but also those with

$$\tilde{f}_n < \text{lb}F(Z_{ni}).$$

Thus: the earlier a small function value can be found the more the computational cost is reduced.

There are many possible techniques for getting lower function values. Here we mention three well-known and useful techniques depending on the differentiability properties of f:

(i) search methods,

(ii) descent methods (steepest descent, conjugate gradients, etc.),

(iii) Newton-like methods.

In order to be complete, one would have to list almost all methods which are used in nonlinear optimization theory. All these methods are local methods: One starts with one point and tries to find a point with a smaller function value.

3. *The interval Newton method.* This is a global method and it provides a reliable technique for enclosing all critical points of f in X. Interval Newton methods can be applied in two different manners:

(i) The method is applied (to f' and X) until all critical points of f are included in sufficiently small boxes Z, say with $w(Z) < \epsilon$ for example. Then the search for the global minimizers may be restricted to the edge of X and the boxes Z. This approach is not so effective as the following:

(ii) Each iteration of the interval Newton method is combined with the monotonicity test and the midpoint test. This means that the basic steps of the algorithms and the acceleration steps merge. This procedure avoids superfluous interval Newton iterations in boxes which are strictly monotone or which have too large function values.

4. *Use of good inclusion functions.* The importance of inclusion functions of a higher order is manifested in Sec. 3.7. Further, if the meanvalue form is used (which is an inclusion function of order 2) then the optimum developing point instead of the midpoint of the boxes should be used. Our tests show a saving of about 10% if that point is used.

5. *Recursive (automatic) differentiation arithmetic* (not to be confused with symbolic differentiation). This technique helps to avoid the large numerical costs when computing derivatives or their inclusion functions, or expressions like $(x - c)^T f'(c), (Y - c) f'(Y), (x - c) f'(Y), (Y - c)^T f'(Y)(Y - c)$, etc. where $x, c \in X$, $Y \in \mathbf{I}(X)$. The recursive differentiation arithmetic is so designed that simultaneously with the computation of $f(x)$ the derivative values or the values of the expressions mentioned above are computed. A detailed description of this fascinating technique is beyond the scope of this book, but see, however, McCormick (1983), who calls this technique *factorable programming*, Rall (1981), (1983), or Kagiwada-Kalaba-Rasakhoo-Spinngarn (1986). In case of non-interval computations we recommend this technique together with updating methods (cf. for example McCormick (1983)) without restrictions. Our own experiences, how-

ever, show that in the case of interval evaluations like $(Y-c)f'(Y)$ or in the case of computing generalized gradients, it is not always wise to use this arithmetic. The reason is that in these cases arithmetic information can be lost and that the widths of the interval values increase unnecessarily, since the recursive differentiation arithmetic does not consider the special arithmetical structure of the formulas for the derivates.

Before we discuss the acceleration devices in detail let us give an overview of the differentiability assumptions which are needed for the devices.

(a) *f continuous:*

Search methods.

(b) *f has a generalized gradient:*

Monotonicity test,
meanvalue form.

(c) *f is continuously differentiable:*

descent and gradient methods,
recursive differentiation arithmetic.

(d) *f is twice continuously differentiable:*

Newton-like methods,
interval Newton methods.

3.12 Acceleration Devices: Detailed Description

Some of the acceleration devices were discussed in Chs. 1 and 2 and some are referred to other sources. In this section the monotonicity test and the devices for finding a lower function value remain to be described. Furthermore, we devote some space to the problem of *where* the devices should be merged among the steps of Alg. 1 to 3.

The monotonicity test

This is a very effective means for discovering whether f is strictly monotone in a subbox $Y \in \mathbf{I}(X)$. If f turns out to be strictly monotone over Y then the interior of Y cannot contain a global minimizer. The edge of Y can contain a global minimizer if that part of the edge which has the smallest function values is also part of the edge of X. Otherwise, no global minimizer lies in Y.

Let $\dfrac{\partial f(x)}{\partial x_i}$ be the i-th components of $f'(x)$, the gradient of f, at x and let $\partial f_i(x)$ be the i-th component of $\partial f(x)$, the generalized gradient of f at x. Let F_i be an inclusion function for $\dfrac{\partial f}{\partial x_i}$ or ∂f_i, that is,

$$\frac{\partial f(x)}{\partial x_i} \in F_i(Y) \text{ or } \partial f_i(x) \in F_i(Y) \text{ for any } x \in Y.$$

Let Y_i and X_i be the i-th components of Y and X. We set $Y_i = [a_i, b_i]$ and $X_i = [c_i, d_i]$. Clearly, if $0 \notin F_i(Y)$ then f is strictly monotone over f with respect to the i-th coordinate direction. This is all that is required for the test.

The monotonicity test consists of 2 parts and it is destined to be applied to the boxes $Y = V_1, V_2$ (cf. Step 7 of Alg. 1 and Step 6 of Alg. 2 and Step 6 of Alg. 3):

Test for strictly monotone increasing:
For some $i = 1, \ldots, m$, if $0 < \mathrm{lb} F_i(Y)$ then

1. *if $c_i < a_i$ then discard the pair (Y, y) from further processing (i.e. it will not be added to the list in later steps),*

2. *if $c_i = a_i$ then set*

$$\begin{aligned} Y' &:= Y_1 \times \ldots \times Y_{i-1} \times a_i \times Y_{i+1} \times \ldots \times Y_m, \\ y' &:= \mathrm{lb} F_i(Y') \end{aligned}$$

and replace (Y, y) by (Y', y') for further processing. This means that $(Y, y) := (Y', y')$ in terms of the algorithm.

Test for strictly monotone decreasing:
For some $i = 1, \ldots, m$, if $\mathrm{ub} F_i(Y) < 0$ then

1. *if $b_i < d_i$ then discard the pair (Y, y) from further processing,*

2. *if $b_i = d_i$ then set*

$$Y' := Y_1 \times \ldots \times Y_{i-1} \times b_i \times Y_{i+1} \times \ldots \times Y_m,$$
$$y' := \mathrm{lb} F_i(Y')$$

and replace (Y, y) by (Y', y') for further processing (which means, set $(Y, y) := (Y', y')$).

Remark. One has to be careful using generalized gradients. They have to be considered with respect to the whole domain of f, that is X, and not only with respect to Y. For example, let $f(x) = |x|$ for $x \in X = [-1, 1]$. If $X = Y_1 \cup Y_2$ where $Y_1 = [-1, 0], Y_2 = [0, 1]$ then one could conclude that $\partial f(x) = -1$ in Y_1 and $\partial f(x) = 1$ in Y_2. Thus, by the monotonicity tests, we can discard Y_1 and Y_2 which implies that also the minimizer, $x^* = 0$, is discarded which gives a wrong result. The right procedure is first to consider

$$\partial f(x) = \begin{cases} -1 & \text{if } x < 0, \\ [-1, 1] & \text{if } x = 0, \\ 1 & \text{if } x > 0. \end{cases}$$

Then $[-1, 1] \subseteq F(Y_i), i = 1, 2$, for any inclusion function F of ∂f in X, and the monotonicity test will not cause any discarding at this stage of the algorithm.

Finding a lower function value

(a) *Nondifferentiable functions:* Since no straight linearization will be available for the objective function f, points with lower function values can best be found with line search via difference procedures. One such simple procedure is given here. Let x_0 be the initial point, for instance, the midpoint of the box Y where the procedure shall be applied. Choose some $\rho > 0$ and compute for each coordinate direction i the values $f(x_0 \mp \rho e_i)$, where e_i is the i-th unit vector of

$\mathbf{R}^m$, provided the arguments remain in X. The box Y may be left since, for the moment, we are only interested in getting low function values and do not need the connections to the box Y where we search. One of the (at most) $2m$ arguments will take the lowest function value, and another search step is started from this point, etc. The search steps should be continued as long as significant lowering of the values is observed. More elaborate versions of this technique are found in Aird-Rice (1977) and in Findler-Lo-Lo (1987).

Locally Lipschitz functions. They have at each point a generalized gradient. Basically, one can use similar methods as in the smooth case, but one has to replace the gradient, cf. next item (b), by a subgradient (which is an element of the generalized gradient) or by a "bundle" of subgradients. Their convergence theory is not yet complete, which does not matter since we do not need convergence but just a few steps for getting lower function values. A deeper discussion would again be beyond the scope of this monograph. We refer to Fletcher (1981), Goffin (1977), Lemaréchal (1980), Mifflin (1977), Shor (1983), Wolfe (1975), Zowe (1985).

(b) $\mathbf{C}^1$ *functions.* The first method is the *steepest descent method* due to Cauchy. Even though better methods exist we cite it because it is widely used, cf. Sec. 1.5. If x_0 is the initial point (midpoint of some box Y) the direction of largest decrease is $-f'(x_0)$. Then the next point is given by

$$x_1 := x_0 - \rho_0 f'(x_0)$$

where ρ_0 is a solution (global or local minimizer) of the one-dimensional optimization problem

$$\min f(x_0 - \rho f'(x_0)) \text{ subject to } \rho \geq 0.$$

If the box X is left by this step then one can reduce the step as is done in the constrained case, cf. Ch. 5. Since the optimization problem mentioned is only of secondary importance, a crude approximation to ρ_0 will do. If the reduction in function value from $f(x_0)$ to $f(x_1)$ is significant then the process should be continued with x_1.

This method has been subsumed by the *conjugate gradient method* of Fletcher-Reeves (1964) where direction information from the previous step is taken into consideration, cf. Sec. 1.5. If the box X is left by the steps, a reduced step width is used as for the constrained problem; see Ch. 5. In special cases, improvements of the Fletcher-Reeves method are available, cf. Powell (1986).

(c) $\mathbf{C^2}$ *functions.* If the objective function f is $\mathbf{C^2}$ then Newton-like methods are commonly favored. The prototype algorithm is described in Sec. 1.5. If the method does not show convergence one should terminate "finding a lower function value" in the box under consideration. When calculating the Hessian, one should not forget that there exist updating procedures and that one might use other facilities for calculating derivatives, cf. Sec. 1.5.

The remaining acceleration devices mentioned have already been discussed in Ch. 2 or they are referred to other sources.

The position of the acceleration devices

Where should the acceleration devices be added to the algorithms we have treated in this chapter? We recommend to insert them after Step 6 of Alg. 1, 2, 3 where the leading box Y is bisected into the subboxes V_1 and V_2. If some or all of the devices are added we suggest the following order (note that the steps of Alg. 1, 2, 3 are changed slightly in order to obtain a better fitting):

(i) For $i = 1, 2$ do

  (a) apply monotonicity test to V_i which results either in the deletion of V_i or in the keeping of V_i, or in the keeping of an edge which is also denoted by V_i,

  (b) determine $v_i = \mathrm{lb}F(V_i)$,

  (c) apply midpoint test to V_i,

  (d) apply one iteration of interval Newton method to V_i getting at most 2 boxes W_{i1}, W_{i2}. Apply monotonicity test to W_{i1}, W_{i2}, determine $w_{ij} = \mathrm{lb}F(W_{ij}), j = 1, 2$, and apply midpoint test to W_{i1}, W_{i2}. Enter the pairs $(W_{ij}, w_{ij}), j = 1, 2$ - if not yet removed - to the list in the right order.

(ii) If the interval Newton method is not used then apply a technique for "Finding a lower function value" to V_1 if $v_1 \leq v_2$, otherwise to V_2 (this means: use mid V_1 or mid V_2 as starting point for the chosen technique).

Remarks. (1) The reason for this ordering is that in most of the cases the meanvalue form for F is used,

$$F(Y) = f(c) + (Y-c)^T F'(Y),$$

where $Y = V_i$, and $c = $ mid Y. This means that first of all $F'(Y) = (F_1(Y), \ldots, F_m(Y))^T$ is computed, where $F'(Y)$ is an inclusion of the gradient of f over Y. Immediately after the computation of $F_i(Y)$ the monotonicity test with respect to the i-th coordinate direction can be applied ($i = 1, \ldots, m$) and Y eventually discarded such that the further processing of Y - for example, the computation of $F(Y)$ - is avoided. If Y is not discarded the values $F(Y)$ and $y = lbF(Y)$ are computed and the midpoint test is next, etc.

(2) If a V_i is discarded then V_i is no longer on the list and it will not be further processed by any devices that might have been incorporated in the algorithm.

(3) If the interval Newton method is incorporated then the last element x_s of the sequence $(x_k)_{k=0}^s$ which is generated by the local iteration procedure shall be used for the updating of $\tilde{f}$ since the chances are good that $f(x_s)$ is near a local minimum. This is also the reason that the boxes W_{ij} are not used for "finding a lower function value" since the local quasi Newton process which generates this sequence is a substitute for a search procedure, cf. Sec. 2.10.

(4) If V_i is discarded by the monotonicity test then that edge (hypercube of one dimension lower than V_i) of V_i which has the smaller function values may also be provided for a "finding a lower function value" procedure after comparison with the other provided box or its edge. See the monotonicity test for the construction of such an edge

Y' of Y. However, only the midpoint of the edge is needed for our purposes.

(5) The "finding of a lower function value" is especially important at the beginning of the computation as discussed earlier. It is however not necessary to apply the related procedures in each iteration and it is quite superfluous at the end of the computation when the monotonicity situation is stable and the boxes concentrate around local minimizers. We suggest the following frequency:

(α) Apply a technique after the starting box X has been bisected, to one of the two parts of X, that is, when list L_2 is built. The continuation depends on the ordering used and shall be described recursively:

(β) In case of Alg. 1 and 2: If a technique has been applied in list L_n, and if l_n is the length of the list, then the technique shall be applied again after l_n iterations, that is, in list L_{n+l_n}.

In case of Alg. 3 which is the important case: Always apply a technique if $Y = \tilde{Y}$ (cf. Alg. 3). The reason is that $\tilde{Y}$ is the box with the smallest lower bound of f, and thus it is likely that $\tilde{Y}$ contains points with low function values.

(γ) If the monotonicity test is used, one may stop looking for smaller function values when the list lengths, l_n, begin to shrink.

3.13 Numerical Examples

We show the application of Hansen's algorithm (Alg. 3) to 3 test functions. The results are mainly taken from Moore-Ratschek (1987). The calculations were done on an Apple IIe microcomputer equipped with a PASCAL-SC software system and on a Honeywell Multics system where interval arithmetic was implemented in FORTRAN. The final representation of the data from the two systems was shortened. Numerical results of Alg. 2 can be found in Ichida-Fujii (1979); further numerical results of Alg. 3 can be found in Hansen (1979), (1980),

Rall (1985), Walster-Hansen-Sengupta (1985). The symbols which are used in the sequel mean:

$X = X_1 \times X_2$ basic domain (Step 1 of Alg. 3)),

ϵ intended maximum box width (termination criterion (3.25)),

ϵ_0 achieved maximum box width,

N number of inclusion function evaluations,

F^* inclusion interval for f^*,

l length (number of boxes) of the final list. Each of these final boxes has width at most $\epsilon_0 \leq \epsilon$,

$E \pm k$ $10^{\pm k}$.

X, F and ϵ are input data; ϵ_0, N, F^* and l are output data. The boxes of the final list are specified later. We used (3.25) as the termination criterion.

Example 1. Six hump camel back function,

$$f(x_1, x_2) = 4x_1^2 - 2.1x_1^4 + \frac{1}{3}x_1^6 + x_1 x_2 - 4x_2^2 + 4x_2^4.$$

The Taylor form of 2nd order, T_2, was taken as an inclusion function for f; see Sec. 2.7. The monotonicity test was used as a device for accelerating the calculations. The data are:

$$\begin{aligned}
X &= [-2.5, 2.5]^2, \\
\epsilon &= E - 6, \\
\epsilon_0 &= 7.16E - 7, \\
N &= 300, \\
F^* &= [-1.03162\ 845353, -1.03162\ 845348], \\
w(F^*) &= 5E - 11, \\
l &= 2.
\end{aligned}$$

The boxes of the final list are:

$$\begin{aligned}
Z_1 &= [-8.984\ 209, -8.984\ 149]E - 2 \times \\
&\quad [7.126\ 557, 7.126\ 565]E - 1, \\
Z_2 &= -Z_1.
\end{aligned}$$

Even though the probability is very high that each of the final boxes contains exactly one global minimizer it cannot be guaranteed. The theory says only that $X^* = \bigcap_{n=1}^{\infty} U_n$, i.e. $X^* \subseteq U_n$ if L_n is the final list before termination. But there is no specific theory of how the global minimizers are distributed to the final boxes, or, how many global minimizers do exist. Thus, one has to use more information in order to analyze the results. This can be done, for instance, in the following manner: In order to prove that each Z_i contains exactly one global minimum point, we consider the interval matrix

$$H_f(Z) = \begin{pmatrix} G(Z) & 1 \\ 1 & -8 + 48W^2 \end{pmatrix}$$

where $Z = V \times W$ and $G(Z) = 8 - 25.2V^2 + V^4$. For simplicity, the components of Z are here denoted by V and W. The matrix $H_f(Z)$ contains the Hessian matrices $H_f(z)$ of f for all $z \in Z$. The principal minors of $H_f(Z)$ for $Z = Z_1, Z_2$ are:

$$G(Z_1) = G(Z_2) = [7.796660,\ 7.796664],$$

and

$$\text{Det } H_f(Z_1) = \text{Det } H_f(Z_2) = [18943.44,\ 18943.45].$$

They are both positive. Therefore f is strictly convex in Z_1 and Z_2, and Z_1 and Z_2 contain *at most* one global minimum point each. Finally, Z_1 and Z_2 do in fact contain a global minimum point each, since $x^* \in Z_1$ implies $-x^* \in Z_2$ and $g(x^*) = g(-x^*)$, and conversely.

In order to demonstrate the dependency of N on the inclusion functions used we present some further results with different inclusion functions but with the same box X and the same required final box width ϵ: We obtained

(i) $N = 326$ when the meanvalue form was used for F,

(ii) $N = 274$ when the meanvalue form with Baumann's optimum developing point was used for F.

Example 2. Branin's function,

$$f(x_1, x_2) = (x_2 - bx_1^2 + cx_1 - d)^2 = e(1 - f)\cos x_1 + e,$$

with
$$b = 5.1/(4\pi^2), c = 5/\pi, d = 6, e = 10, f = 1/(8\pi).$$

The Taylor-form of second order was again used as an inclusion function for f. Alg. 3 was supplemented by the monotonicity test as in Example 1. The following results were obtained:

$$\begin{aligned}
X &= [-5, 10] \times [0, 15], \\
\epsilon &= 1E-6, \\
\epsilon_0 &= 8.941E-7, \\
N &= 1166, \\
F^* &= [3.97887\ 357712, 3.97887\ 357770]E-1, \\
w(F^*) &= 5.8E-11, \\
l &= 18.
\end{aligned}$$

The resulting 18 final boxes, Z_i, are not shown here. They are instead grouped together and included in three boxes $\tilde{Z}_1$ as small as possible:

$$\begin{aligned}
\tilde{Z}_1 &= [-3.14159\ 036, -3.14159\ 393] \times [12.274\ 9936, 12.275\ 0035], \\
\tilde{Z}_2 &= [3.14159\ 15, 3.14159\ 34] \times [2.274\ 9987, 2.275\ 0015], \\
\tilde{Z}_3 &= [9.4247\ 703, 9.4247\ 882] \times [2.474\ 998, 2.475\ 002].
\end{aligned}$$

Each of the 3 boxes $\tilde{Z}_i$ contains exactly one global minimizer. This knowledge however must be taken from other sources even if it is indicated by the numerical data.

We again used different inclusion functions and obtained:

(i) $N = 1354$ when the meanvalue form was used for F,

(ii) $N = 1194$ when the meanvalue form with Baumann's optimum developing point was used for F.

Example 3. The function

$$\begin{aligned}
f(x_1, x_2) &= x_1 \sin(1/x_1) + x_2 \mid x_1 \mid, & \text{if } x_1 \neq 0, \\
&= 0 & \text{if } x_1 = 0,
\end{aligned}$$

was already discussed in Sec. 3.5. It has infinitely many global minimum points in the domain

$$X = [0, 1] \times [1, 1+E-8].$$

Numerical Examples

The same inclusion function as in Sec. 3.5 was used. The computations were terminated by an emergency criterion for avoiding stack overflow (the list became too long). The maximum box width

$$\epsilon_0 = 3.81E - 6$$

was attained. The other results were:

$$\begin{aligned} N &= 2198, \\ F^* &= [0,\ 2.75]E - 10, \\ l &= 522. \end{aligned}$$

It is neither possible nor worthwhile to print the 522 boxes of the final list. Many of them are connected. It may be of interest that the area of the union of the boxes is about 0.2% of the area of the starting domain X.

Chapter 4

Unconstrained Optimization over Unbounded Domains

4.1 Introduction

Almost all of the well-known methods for solving the global unconstrained optimization problem involve a bounded subdomain $X \subseteq \mathbf{R}^m$, for the function to which the method is applied. Primarily, this is for computational reasons. Secondarily, the theoretical investigations are simplified, because of the existence of accumulation points of a sequence in a compact domain and because of other compactness arguments: see proof of Theorem 4 of Ch. 3. Therefore, an appropriate bounded area X must be known *a priori* or must be determined by means of an analysis of the problem. If this is not possible, linear substitutions such as $x = 1/s$ are commonly used to transform unbounded parts of the domain into bounded areas. These substitutions are however rather troublesome to program because of the many cases which may arise. Let, for example, $f(x,y) = xy$ for $x, y \in \mathbf{R}$. The plain substitution is $x = 1/s, y = 1/t$. This results in the following cases where a fixed or variable bound $k > 0$ has to be chosen:

(a) $|x|, |y| \le k$,

(b) $|x| \le k < |y|$,

(c) $|y| \le k < |x|$,

(d) $k < |x|, |y|$.

Let us consider case (b) for the moment. The necessary substitution $y = 1/t$ leads to two disjoint areas, i.e. $y < -k$ and $y > k$ which are maintained as separate areas during the computation. Symmetry arguments are not applicable in general. Similar considerations lead to two disjoint areas in case (c) and to four different areas in case (d). Finally the function values may also tend to infinity as for example in case (b) where $f(x,y) = x/t$. Thus the function needs to be transformed, $g(x,y) = 1/f(x,y) = t/x$. But what happens when x is near zero? The list of awkward points could be continued. Certainly, the situation above is outlined in an exaggerated fashion; however, there is no guarantee that it can be avoided.

The technique we provide in this chapter is due to Ratschek-Voller (1988) and it avoids such tortuous paths. It is destined to solve the global minimization problem for a continuous function, $f : \mathbf{R}^m \to \mathbf{R}$. The search for the global minimizers is accomplished in the whole domain, $\mathbf{R}^m$. Bounds of the global minimum, f^*, are generated, and one or several boxes of prescribed size which include all global minimizers are produced. It can be detected when f has no global minimum at all, and further, whether f is bounded from below or not. Examples of such functions are $f(x) = e^x$ and $f(x) = x$. If the technique is applied on a computer, the sharpness of these detections is limited by the finite number representation of computers. The detections - although weakened - remain logically valid in this case such that the user can trust them.

The technique that we provide to cover the whole space $\mathbf{R}^m$ when looking for global minimizers is best demonstrated by applying it to Alg. 3 of Ch. 3. Similar algorithms such as Alg. 1 of Ch. 3 or Alg. 2 of Ch. 3 as well as the algorithms for solving the constrained problem as they are developed in the sequel may also be used. The common feature of these algorithms is that they are based on the branch and bound principle and that the bounds are determined by means of interval arithmetic tools.

In Sec. 4.2 and Sec. 4.3, the optimization problem and Alg. 3 of Ch. 3 are extended and admitted to functions which are defined on $\mathbf{R}^m$. For this reason, a compactification of the space $\mathbf{R}^m$ is intro-

Introduction

duced, $\overline{\mathbf{R}^m} := (\overline{\mathbf{R}})^m$ where $\overline{\mathbf{R}} := \mathbf{R} \cup \{\infty, -\infty\}$. The advantages of this compactification are threefold: (i) The investigation of the convergence properties can make use of compactness principles. (ii) The interpretation in $\mathbf{R}^m$ of the results obtained in $\overline{\mathbf{R}^m}$ is straightforward and there is no need to distinguish between the bounded and the unbounded case. (iii) The step from the exact execution of the algorithm in $\overline{\mathbf{R}^m}$ to its numerical execution on a computer is small because the latter operates in $[-L, L]^m$ where L denotes the largest representable number of the machine under consideration. $\overline{\mathbf{R}^m}$ and $[-L, L]^m$ are topologically very similar.

In Sec. 4.4, the monotonicity test is carried over to the unbounded case. The usual assumption for f to be differentiable can be weakened. It is only necessary that, at any point, a generalized gradient which can also be unbounded exists in order to apply the test.

In Sec. 4.5 it is shown practically how to get the tools - such as inclusion functions - which are required for the extended algorithm, i.e. an arithmetic for unbounded noncompact intervals is introduced. For several reasons, it is different from Kahan's (1968) or Laveuve's (1975) arithmetic. As a consequence, Moore's principle of natural interval extension (cf. Sec. 2.6) can be recursively defined for programmable functions over unbounded domains. This is important for getting the inclusion functions mentioned which generate bounds for the objective function, f, over any subbox $Y \subseteq \mathbf{R}^m$.

In Sec. 4.6, the relationships between the numerical and the exact realization of the extended algorithm are discussed.

In Sec. 4.7, numerical examples show that the practical computation involves no difficulties at all.

Comparing Sections 4.2, 4.5, and 4.6, we are faced with 3 kinds of infinite intervals:

(i) *Compactified unbounded intervals* such as $[a, \infty] \subseteq \overline{\mathbf{R}}$, cf. Sec. 4.2. They are needed both for the execution of the algorithm and for the discussion of its convergence properties. Since only topological arguments and no arithmetic are used for this discussion, no arithmetic need be defined for compactified unbounded intervals.

(ii) *Unbounded noncompactified intervals* such as $[a, \infty) \subseteq \mathbf{R}$, cf. Sec. 4.5. They occur when the bounds of f over unbounded subdomains are determined. Thus, an arithmetic for such intervals is defined.

(iii) Both kinds of intervals mentioned in (i) and (ii) must be approximated by *machine intervals* such as $[a, L]$ when computing on a machine, cf. Sec. 4.6. Although they look like bounded intervals, they are of unbounded character since L represents the numbers from L to ∞.

4.2 The Algorithm over Unbounded Domains

We will first provide a compactification $\overline{\mathbf{R}^m}$ of $\mathbf{R}^m$. Then Alg. 3 of Ch. 3 and some convergence theorems of Ch. 3 will be extended to the compactified case immediately. The results gained in $\overline{\mathbf{R}^m}$ can then be converted into the originally required results in $\mathbf{R}^m$. The compactification could be avoided but it simplifies matters considerably.

In order to apply Alg. 3 of Ch. 3 to $\overline{\mathbf{R}^m}$, the midpoint and the width of boxes in $\overline{\mathbf{R}^m}$ must be defined, and, further, the given function, f, and its inclusion function, F, must be extended. This requires some notation.

Let $\overline{\mathbf{R}} := \mathbf{R} \cup \{-\infty, \infty\} = [-\infty, \infty]$ be the two-point compactification of $\mathbf{R}$ and $\overline{\mathbf{R}^m} := \overline{\mathbf{R}}^m$ be the m-fold topological product of $\overline{\mathbf{R}}$. If $A \subseteq \mathbf{R}^m$ or $A \subseteq \overline{\mathbf{R}^m}$, we denote the compact hull of A with respect to this compactification by $\overline{A}$. It is not necessary to describe this topology in detail. It will be used just for the concept of a compact interval or box and for the convergence concept.

Let $\overline{\mathbf{I}}$ be the set of compact intervals of $\overline{\mathbf{R}}$. Thus, $\overline{\mathbf{I}}$ consists of intervals $[a, b]$, $[a, \infty]$, $[-\infty, b]$, where $a, b \in \mathbf{R}$, and of the point intervals $-\infty = [-\infty, -\infty]$, $\infty = [\infty, \infty]$. A box $Y = (Y_1, \ldots, Y_m) \subseteq \overline{\mathbf{R}^m}$ is compact iff $Y_1, \ldots, Y_m \in \overline{\mathbf{I}}$, by Tychnoff's theorem in topology.

When does a sequence (x_n) of $\overline{\mathbf{R}}$ converge to a point $x \in \overline{\mathbf{R}}$? There are two cases: If $x \in \mathbf{R}$ then $x_n \to x$ iff an index x_0 exists such that the sequence $(x_n)_{n\geq n_0}$ converges to x (with respect to the natural topology of $\mathbf{R}$). If $x = \infty$ (resp. $x = -\infty$) then $x_n \to x$ iff to any given real number $k > 0$ an index n_0 exists such that $x_n \geq k$ (resp. $x_n \leq -k$) for all $n \geq n_0$. Let now $x_n = (x_n^1, \ldots, x_n^m) \in \overline{\mathbf{B}^m}$ and $x = (x^1, \ldots, x^m)$. Then owing to topology, $x_n \to x$ iff $x_n^i \to x^i$ for $i = 1, \ldots, m$.

A point $x \in \overline{\mathbf{R}^m}$ is an accumulation point of a sequence (x_n) of $\overline{\mathbf{R}^m}$ iff a subsequence of (x_n) converges to x. If (x_n) is a sequence in $\overline{\mathbf{R}}$ then
$$\liminf_{n\to\infty} x_n$$
is the smallest accumulation point of (x_n).

Let $\mathbf{I}_\infty$ be the set of all closed (but not necessarily bounded) intervals of $\mathbf{R}$. Thus, the intervals $[a, b], [a, \infty), (-\infty, b]$, and $(-\infty, \infty) = \mathbf{R}$ belong to $\mathbf{I}_\infty$ where $a, b \in \mathbf{R}$.

Let $A \subseteq \overline{\mathbf{R}^m}$, then $\mathbf{I}_\infty(A) := \{Y \in \mathbf{I}_\infty^m : Y \subseteq A\}$ and $\overline{\mathbf{I}}(A) := \{Y \in \overline{\mathbf{I}}^m : Y \subseteq A\}$. We note that $\mathbf{I}_\infty^m = \mathbf{I}_\infty(\mathbf{R}^m) = \mathbf{I}_\infty(\overline{\mathbf{R}^m})$ and $\overline{\mathbf{I}}^m = \overline{\mathbf{I}}(\overline{\mathbf{R}^m})$. Further, $\mathbf{I}(A) := \{Y \in \mathbf{I}^m : Y \subseteq A\}$.

Let $A \in \mathbf{I}_\infty^m$ and $f : A \to \mathbf{R}$ be given. We want to extend f to a function $f_o : \overline{A} \to \overline{\mathbf{R}} :$ Let $x \in \overline{A}$ then
$$f_o(x) := \min\{\liminf_{n\to\infty} f(x_n) : x_n \in A, x_n \to x\} \qquad (4.1)$$
where the convergence of the sequences (x_n) to x is subject to the topology of $\overline{\mathbf{R}^m}$. Note that f_o need not be continuous, even when f is.

For example, if $f(x) = e^x$, $x \in \mathbf{R}$, then $f_o(-\infty) = 0$ and $f_o(\infty) = \infty$. If $f(x) = \sin x, x \in \mathbf{R}$, then $f_o(-\infty) = f_o(\infty) = -1$.

Analogously, if $F : \mathbf{I}(A) \to \mathbf{I}$ is an inclusion function for f, we want to extend F to an inclusion function $F_o : \overline{\mathbf{I}}(\overline{A}) \to \overline{\mathbf{I}}$ for f_o, that is,
$$\Box f_o(\overline{Y}) \subseteq F_o(\overline{Y}) \text{ for any } Y \in \mathbf{I}_\infty(A). \qquad (4.2)$$
We do not need inclusions of f over boxes $Z = Z_1 \times \ldots \times Z_m$ where any component Z_i is just ∞ or $-\infty$. Such cases are not considered in (4.2) and also in the following definition (4.3) which simplifies matters. Thus we define F_o by
$$F_o(\overline{Y}) := \overline{F(Y)} \text{ for } Y \in \mathbf{I}_\infty(A). \qquad (4.3)$$

We call F and also F_o *non-wasteful* if, given any $Y \in \mathbf{I}_\infty(A)$, a partition of A into bounded boxes exists, $A = \cup_{i \in J} B_i$, with $B_i \in \mathbf{I}(A)$ and some index set J, such that

$$F(Y) \subseteq \overline{\cup_{i \in J} F(B_i)}. \tag{4.4}$$

(4.4) is necessary for getting reasonable convergence properties. It is a very natural condition, since each programmer would automatically construct non-wasteful inclusions.

For example, let $f(x) = x_1^2 + x_2$, $x \in \mathbf{R}^2$, then $F(Y) = Y_1^2 + Y_2$, $Y \in \mathbf{I}_\infty^2$, is non-wasteful. If $f(x) = \sin x$, $x \in \mathbf{R}$, $F(Y) = [-1, 1]$, if $Y \in \mathbf{I}$, and $F(Y) = [-2, 2]$, if $Y \in \mathbf{I}_\infty \setminus \mathbf{I}$, then F is wasteful. That the compact hull in (4.4) is necessary is best made clear by another example: Let $A = (-\infty, 0]$, and $f : A \to \mathbf{R}$ be defined by $f(x) = e^x$. If $Y = (-\infty, a], a < 0$, then $\Box f(Y) = (0, e^a]$ is not an interval belonging to $\mathbf{I}$ or $\mathbf{I}_\infty$. This fact has to be considered when an inclusion function is constructed: Let $F(Y) := \overline{\Box f(Y)}$ for $Y \in \mathbf{I}_\infty(A)$ then F is non-wasteful but $F(Y) \subseteq \cup_{i \in J} F(B_i)$ does not hold for any partition provided.

Since there is no danger of misunderstanding we also write f and F instead of f_o and F_o in the sequel.

Using the notation just introduced, the *global unconstrained optimization problem over unbounded domains* can be written down concisely as follows:

Let $X \in \mathbf{I}_\infty^m$ and $f : X \to \mathbf{R}$ be continuous. The problem to be solved is

$$\min_{x \in X} f(x). \tag{4.5}$$

This means that we expect the program to find out whether the minimum exists or not. If the minimum does not exist we want to know whether f is unbounded from below or not. The unbounded problem, (4.5), is first reduced to the *compactified problem*, which is

$$\min_{x \in \overline{X}} f(x). \tag{4.6}$$

This minimum always exists if the extension of f on $\overline{X}$ is defined via (4.1).

The Algorithm over Unbounded Domains

It is now necessary to introduce a width for unbounded boxes in order to define the global optimization algorithm for unbounded domains. This width may be defined in a variety of ways resulting in formulas of greater or lesser complexity. We do not expect our formula to be either elegant or of theoretical interest (like the chordal-distance on the Riemann sphere); however, it must be simple and appropriate for our purposes. Our formula depends on a global parameter λ which the user or programmer may choose, such that the global minimizers are suspected to lie in the box $[-\lambda, \lambda]^m$. This choice of λ which will also influence the bisection process, forces areas outside of $[-\lambda, \lambda]^m$ to be cut off as soon as possible. If the user has no conjecture as to the location of the minimizers and if he does not assign a value to λ, then our program version sets $\lambda := 10$. If the choice of the user is wrong, the program is still correct but it is then slower.

The *width* of an unbounded box $Y = Y_1 \times \ldots \times Y_m \in \overline{\mathbf{I}}^m$ is defined as follows. Let $0 < \lambda < \infty$ and $a \in \mathbf{R}$. Then

$$w([a, \infty]) := \begin{cases} \lambda^2/a & \text{if } a \geq 10^{-10}, \\ \max(1, \lambda^2) & \text{otherwise}, \end{cases}$$

$$w([-\infty, a]) := w([-a, \infty]),$$

$$w([-\infty, \infty]) := \lambda^2 10^{11}, \quad w(\pm[\infty, \infty]) = 0,$$

$$w(Y) := \max_{i=1,\ldots,n} w(Y_i).$$

The bisection cuts the boxes through the midpoint. According to our intention not to dissect $[-\lambda, \lambda]^m$ too early, the *midpoint* of unbounded boxes will also be made dependent on λ. Let $a \in \overline{\mathbf{R}}$ and $Y \in \overline{\mathbf{I}}^m$. Then we set

$$\text{mid}\,[a, \infty] := \begin{cases} \lambda & \text{if } a \leq \lambda, \\ 2a & \text{if } \lambda < a \leq 3\lambda, \\ 10a & \text{if } 3\lambda < a, \end{cases}$$

$$\text{mid}\,[-\infty, a] := -\text{mid}\,[-a, \infty]$$

$$\text{mid}\,Y := (\text{mid}\,Y_1, \ldots, \text{mid}\,Y_m)^T.$$

If, for example, $\lambda = 10$, then the interval $[30, \infty]$ is bisected into $[30, 60]$ and $[60, \infty]$.

We are now ready to establish the prototype algorithm for unbounded domains as follows:

ALGORITHM 1 *shall be syntactically equal to Alg. 3 of Ch. 3, but now, unbounded compactified boxes $\overline{X}$ of $X \in \mathbf{I}_\infty^m$, and inclusion functions $F : \overline{\mathbf{I}}(\overline{X}) \to \overline{\mathbf{I}}$ of the functions $f : \overline{X} \to \overline{\mathbf{R}}$ are admitted as input data. We use the formulas for width and midpoint of unbounded compact boxes as they have just been introduced.*

Algorithm 1 aims to determine the global minimum, f^+, and X^+, the set of global minimizers, of f over $\overline{X}$.

Alg. 1 like Alg. 3 of Ch. 3 produces, at the n-th iteration, a list $\mathbf{L}_n$ consisting of pairs $(Z_{ni}, z_{ni}), i = 1, \ldots, l_n$, where l_n is the list length and $z_{ni} = \mathrm{lb}F(Z_{ni})$. The leading pair of $\mathbf{L}_n$ is denoted by (Y_n, y_n), and $(\tilde{Y}_n, \tilde{y}_n)$ denotes a pair of $\mathbf{L}_n$ satisfying $\tilde{y}_n \leq z_{ni}$ for $i = 1, \ldots, l_n$. The function value $f_n \in \overline{\mathbf{R}}$ is the lowest value of f produced up to the n-th iteration. As before U_n denotes the union of all boxes of $\mathbf{L}_n$, i.e. $\cup_{i=1}^{l_n} Z_{ni}$, and again (U_n) is a nested sequence.

4.3 Convergence Properties

In this section, practically reasonable assumptions are looked for under which Alg. 1 converges to the solution set of the compactified problem, (4.6). This solution is used to derive the originally required solution, of (4.5). Termination criteria are touched upon briefly.

First of all,
$$w(Y_n) \to 0 \text{ as } n \to \infty. \tag{4.7}$$
The proof is similar to the proof for the bounded case in Sec. 3.4 and is suppressed. Also, from the execution of Alg. 1, we have
$$\tilde{y}_n \leq f^+ \leq f_n \text{ for all } n. \tag{4.8}$$
In order to describe the convergence of U_n to X^+ we extend the Hausdorff-distance, d, as it is introduced in Sec. 3.2, to compact subsets of $\overline{\mathbf{R}^m}$ and denote this extension by d as well. Let $a = (a_1, \ldots, a_m) \in \overline{\mathbf{R}^m}$, and A, B be compact subsets of $\overline{\mathbf{R}^m}$. We define

$$d_o(a, B) := \min_{b \in B}(\sum_{i=1}^m w([\min(a_i, b_i), \max(a_i, b_i)])^2)^{\frac{1}{2}},$$
$$d_o(A, B) := \max_{a \in A} d_o(a, B),$$
$$d(A, B) := \max\{d_o(A, B), d_o(B, A)\}.$$

The above formula should not astonish the reader. It is simply a means for making the convergence mentioned precise and controllable. We write $A_n \to B$ instead of $d(A_n, B) \to 0$ in the sequel. Clearly, if $d(A, B) = 0$ then $A = B$.

Let again $X \in \mathbf{I}_\infty^m$, $f : \overline{X} \to \overline{\mathbf{R}}$ and $F : \overline{\mathbf{I}}(\overline{X}) \to \overline{\mathbf{I}}$ be an inclusion function of f. If Alg. 1 is applied to $\overline{X}$ and F then the behavior and the output are completely described by:

If
$$w(F(\overline{Y})) \to 0 \text{ as } w(\overline{Y}) \to 0, \text{ for } Y \in \mathbf{I}_\infty(X), \tag{4.9}$$

then $f_n - \tilde{y}_n \to 0$ and $U_n \to X^+$ as $n \to \infty$. This proposition is true because the corresponding proposition in the bounded case, cf. Ch. 3, depends on compactness and a few properties of the width, and the bounded case can thus be carried straight over to the compactified case.

The assumption (4.9) is, however, too restrictive. Let, for example, $f(x) = \cos x$ and $F(Y) = \Box f(Y)$. Owing to (4.1), (4.3), the extension to $\overline{\mathbf{R}}$ gives $f(\pm\infty) = -1, F(\overline{Y}) = [-1, 1]$ if $\pm\infty \in \overline{Y}$. The global minimum of f over $\overline{\mathbf{R}}$ is $f^+ = -1$, the global minimizers are $\pi + 2k\pi (k = 0, \pm 1, \ldots)$ and $\infty, -\infty$, but condition (4.9) is not satisfied.

Fortunately, assumption (4.9) can be weakened to

$$w(F(Y)) \to 0 \text{ as } w(Y) \to 0, \text{ for } Y \in \mathbf{I}(X), \tag{4.10}$$

if non-wasteful inclusion functions are used. Then again only bounded boxes Y are involved. (4.10) holds for the function F of the example just mentioned.

Let $X \in \mathbf{I}_\infty^m$, let $f : \overline{X} \to \overline{\mathbf{R}}$ satisfy (4.1) and let $F : \overline{\mathbf{I}}(\overline{X}) \to \overline{\mathbf{I}}$ be a non-wasteful inclusion function of f satisfying (4.3). Then (4.6) has a solution, i.e. $X^+ \neq \emptyset$, and f^+ exists. If Alg. 1 is applied to F and $\overline{X}$, then the following holds for the output data:

THEOREM 1 *If (4.10) holds then, as $n \to \infty$,*

(i) $f^+ - \tilde{y}_n \to 0$ as well as $f^+ - y_n \to 0$,

(ii) $f_n - f^+ \to 0$,

(iii) $U_n \to X^+$.

Proof. (i) We show that $f^+ - \tilde{y}_n \to 0$. The second assertion, $f^+ - y_n \to 0$, will follow from (ii), since $\tilde{y}_n \le y_n \le f_n$ holds for any n.
If $f^+ = -\infty$ then $\tilde{y}_n = f^+$ because of (4.8). Let $f^+ \in \mathbf{R}$. We focus on the n-th iteration for a moment. Since F is non-wasteful, we have

$$F(\tilde{Y}_n) = \overline{\cup_{i \in J_n} F(B_{ni})}$$

for some partition $\tilde{Y}_n \cap \mathbf{R}^m = \cup_{i \in J_n} B_{ni}$. Since $\tilde{y}_n = \mathrm{lb} F(\tilde{Y}_n)$ and since $f^+ \in F(B_{ni})$ or $f^+ < \mathrm{lb} F(B_{ni})$ for any $i \in J_n$, the following choice is possible:

If $\tilde{y}_n = -\infty$ we choose $B_{ni_n}, i_n \in J_n$, such that

$$\mathrm{lb} F(B_{ni_n}) < f^+ - 2^n. \qquad (4.11)$$

If $\tilde{y}_n \in \mathbf{R}$ we choose $B_{ni_n}, i_n \in J_n$, such that either $\tilde{y}_n \in F(B_{ni_n})$ or both $\mathrm{lb} F(B_{ni_n}) - \tilde{y}_n < 2^{-n}$ and $|\mathrm{lb} F(B_{ni_n}) - f^+| < 2^{-n}$ hold.

Now, (4.7) implies $w(B_{ni_n}) \to 0$. Since $B_{ni_n} \in \mathbf{I}(X)$, we can apply assumption (4.10) and get $w(F(B_{ni_n})) \to 0$. Comparing this property with (4.11) we see that $\tilde{y}_n \in \mathbf{R}$ holds for sufficiently large n, which means that $\tilde{y}_n \to f^+$.

(ii) We have to show that $f_n \searrow f^+$. Let us, however, assume that $f_n \searrow \alpha$ for some $\alpha > f^+$. Since $f^+ = f(x^+)$ for some $x^+ \in X^+$ and since f satisfies (4.1) there exists a sequence (ξ_k), $\xi_k \in X \subseteq \mathbf{R}^m$ such that

$$\xi_k \to x^+ \text{ and } f(\xi_k) \to f(x^+) \text{ as } k \to \infty.$$

Let k be fixed such that $f(\xi_k) < \alpha$. There exists a sequence (Z'_n) with $\xi_k \in Z'_n$ and Z'_n belonging to $\mathbf{L}_n$, for any n. The existence of (Z'_n) is guaranteed since ξ_k is never excluded by the midpoint test. (4.7) implies $w(Z'_n) \to 0$, and further $Z'_n \to \xi_k$. Thus, $Z'_n \in \mathbf{I}(X)$, for sufficiently large n. We can apply (4.10) and get $w(F(Z'_n)) \to 0$ and $F(Z'_n) \to f(\xi_k)$. For some large n we thus have $\mathrm{ub} F(Z'_n) < \alpha$. Since $f_n \le \mathrm{ub} F(Z'_n)$, we have a contradiction.

(iii) The assertion is proven when we have shown that $X^+ \subseteq U_n$ for any n and that $\cap_{n=1}^\infty U_n \subseteq X^+$. According to the construction of Alg. 1, $X^+ \subseteq U_n$ is obvious. It remains to show that

$$x \in U_n, \text{ for all } n, \text{ implies } x \in X^+.$$

Convergence Properties

We assume that $f(x) > f^+$ in order to get a contradiction. Since x occurs in every list, a sequence (Z'_n) exists where $x \in Z'_n$ and Z'_n belongs to L_n. Since F is non-wasteful a partition of $Z'_n \cap \mathbf{R}^m$ exists, $Z'_n \cap \mathbf{R}^m = \cup_{i \in J_n} B_{ni}$, corresponding to the definition of a non-wasteful inclusion function. If $Z'_n \in \mathbf{I}(X)$, set $Z'_n = B_n = B_{ni_n}$. Otherwise choose $i_n \in J_n$ such that $\mathrm{lb}F(B_{ni_n}) \leq \mathrm{lb}F(Z'_n)$ or that $\mathrm{lb}F(B_{ni_n})$ is asymptotically close to $\mathrm{lb}F(Z'_n)$.

Since $w(Z'_n) \to 0$, it follows that $Z'_n \to x$ and thus $B_{ni_n} \to x$. Let $\xi_n \in B_{ni_n}$ then $f(\xi_n) \to \alpha$ for some $\alpha \geq f(x)$ due to (4.1). $w(B_{ni_n}) \to 0$ implies $w(F(B_{ni_n})) \to 0$ and further, $F(B_{ni_n}) \to \alpha$ and $\mathrm{lb}F(B_{ni_n}) \to \alpha$. From the choice of B_{ni_n} it follows that also $\mathrm{lb}F(Z'_n) \to \alpha$. By (ii), we have $f_n < f(x) \leq \alpha$ for large n such that Z'_n and thus x is discarded by the midpoint test. This gives the contradiction. $\square$

Remark. It is important that the midpoint test is done via the leading boxes, Y_n, and not via $\tilde{Y}_n$ which may be promising since $\tilde{Y}_n$ delivers the lowest lower bound, $\tilde{y}_n$. But, when doing so, parts of Theorem 1 cannot be proven.

Condition (4.10) is too restrictive for functions which have unbounded ranges. Let, for example, $f(x) = x^2$ and $F(Y) = Y^2$ be an inclusion function. Although F is an optimum inclusion function because of $F(Y) = \square f(Y)$ and although the assertions (i) to (iii) of Theorem 1 are satisfied, condition (4.10) does not hold. In such cases the condition

$$w(F(Y)) - w(\square f(Y)) \to 0 \text{ as } w(Y) \to 0,$$

for $Y \in \mathbf{I}(X)$, could be appropriate. In practice, however, this condition is too difficult to verify. Therefore we establish a condition with which we have obtained the best practical results. It combines theoretical as well as practical requirements where the computer may verify the latter for us.

Let $X \in \mathbf{I}_\infty^m$, $f : \overline{X} \to \overline{\mathbf{R}}$, and an inclusion function $F : \overline{\mathbf{I}}(\overline{X}) \to \overline{\mathbf{I}}$ of f be given. We assume that for any given $Z \in \mathbf{I}(X)$

$$w(F(Y)) \to 0 \text{ as } w(Y) \to 0 \text{ for } Y \in \mathbf{I}(Z). \quad (4.12)$$

(Condition (4.12) is very general and not at all restrictive. If, for instance, f is continuous and programmable and if F is constructed via natural interval extensions (cf. Sec. 4.5) then (4.12) is already satisfied.) Then (4.6) has a solution, i.e. $X^+ \neq \emptyset$, and f^+ exists. If Alg. 1 is applied to F and $\overline{X}$ then the following holds for the output data:

THEOREM 2 *If (4.12) holds and if there exists a number n such that the list L_n contains only bounded boxes, that is, boxes of I^m, then the propositions (i), (ii), and (iii) of Theorem 1 are valid.*

Proof. Let Z be the smallest box of I^m which contains the boxes of L_n. That means $X^+ \subseteq U_n \subseteq Z$, due to Alg. 1. We can now think of L_n as a list created by applying Alg. 3 of Ch. 3 to f, F, and Z, such that the assertion of the theorem follows from the results of Ch. 3. $\square$

Example. Let us consider the well-known six hump camel back function,

$$f(x) = 4x_1^2 - 2.1x_1^4 + x_1^6/3 + x_1 x_2 - 4x_2^2 + 4x_2^4, \quad x \in X = \mathbf{R}^2.$$

We compare two inclusion functions, F and F_1, for f over $\overline{X}$,

$$\left.\begin{array}{l} F(Y) := 4Y_1^2 + Y_1^4(Y_1^2/3 - 2.1) + Y_1 Y_2 + 4Y_2^2(Y_2^2 - 1), \\ F_1(Y) := 4Y_1^2 - 2.1Y_1^4 + Y_1^6/3 + Y_1 Y_2 - 4Y_2^2 + 4Y_2^4, \\ F(\overline{Y}) := \overline{F(Y)}, \quad F_1(\overline{Y}) := \overline{F_1(Y)}. \end{array}\right\} \text{ for } Y \in \mathbf{I}_\infty(X).$$

Neither F nor F_1 satisfies (4.10). Both F and F_1 satisfy (4.12). If Alg. 1 is applied to f (using (4.1)), $\overline{X}$ and F then, after a few iterations, the lists L_n do not contain any unbounded boxes. This is not the case if F_1 is used instead of F. The assumptions of Theorem 2 are therefore computationally verified for F, but not for F_1.

Remarks. (1) It is difficult to present precise conditions for inclusion functions to satisfy the bounded box assumption of Theorem 2. In practice, F satisfied this assumption if F was constructed using the following recipe: Let F be so that $-\infty \notin F(\overline{Y})$ for all boxes

$Y = Y_1 \times \ldots \times Y_m \in \mathbf{I}_\infty(X)$, where $Y_i = (-\infty, -a]$ or $Y_i = [a, \infty)$, for some arbitrarily large real $a > 0$.

(2) Alg. 1 might work well even for problems having convergence properties which are not covered by any convergence theorem, which is mainly the case if f is not continuous. The reason for this behavior is that the unions, U_n, always form a nested sequence, where $X^+ \subseteq U_n$, and that $(f_n - \tilde{y}_n)$ is *always* a monotonically decreasing sequence where $\tilde{y}_n \leq f^+ \leq f_n$. It is not too unlikely that U_n or $f_n - \tilde{y}_n$ will be small enough for some reasonable n to establish reasonable solutions of problem (4.6).

Termination criteria. Theorems 1 and 2 suggest three kinds of termination criteria. They shall only be dealt with superficially since they are almost the same as in the bounded case, cf. Sect. 3.10.

Criterion A: Terminate when $f_n - \tilde{y}_n < \epsilon$. This gives an approximation of f^+ and an error estimate.

Criterion B: Terminate when $\lambda(U_n) < \epsilon$, where λ is any "measure" of $\overline{\mathbf{R}}^m$. For example, $\lambda(Y) := w(Y_1) \times \ldots \times w(Y_m)$ if $Y = Y_1 \times \ldots \times Y_m \in \overline{\mathbf{I}}^m$ and $\lambda(U_n) := \sum_{i=1}^{l_n} \lambda(Z_{ni})$, if $U_n = \cup_{i=1}^{l_n} Z_{ni}$. This criterion only works if the measure of X^+ is 0.

Criterion C: Terminate when $w(Z_{ni}) < \epsilon$ for all Z_{ni} of L_n.

Let us return to the problem (4.5) originally posed. Its solution can easily be derived from the solution of the compactified problem, (4.6). The solutions of (4.6) are denoted by X^+, f^+, and the solutions of (4.5) by X^*, f^* if they exist. Therefore, Alg. 1 is appropriate to solve (4.5) via (4.6) if the following theorem is used:

THEOREM 3 *(i) If $f^+ = -\infty$ then (4.5) has no solution, and f is unbounded from below in X.*
(ii) If $f^+ \in \mathbf{R}$ and if $X^* := X^+ \cap \mathbf{R}^m$ is non-empty then X^* and $f^* = f^+$ is the solution of (4.5). If $X^* = \emptyset$ then (4.5) has no solution

but f is bounded from below, and $f^+ = \liminf\limits_{x_n \to x^+} f(x_n)$, $x_n \in X$ and $x^+ \in X^+$.

Proof. Obvious. □

Example. Case (i): $f(x) = x$ will do it. Case (ii): $f(x) = \cos x$ has solutions $X^* = \{(2k+1)\pi : k = 0, \pm 1, \ldots\}$, $X^+ = X^* \cup \{\infty, -\infty\}$ and $f^* = f^+ = -1$. Further $f(x) = e^x$ has no solution w.r.t. (4.5), but $X^+ = \{-\infty\}$ and $f^+ = 0$.

4.4 The Monotonicity Test

The monotonicity test, cf. Sec. 3.12, is a very effective tool for discarding boxes from the lists of interval optimization algorithms. It is even more important in the unbounded case, as functions defined on unbounded domains are frequently strictly monotone (with respect to some coordinate direction) for large values of the variables - like polynomials. Such areas contain no minimizers and can be discarded from the lists. The unbounded case is thus reduced to the bounded one. Therefore we recommend that the monotonicity test be incorporated in the algorithm whenever possible. This test is best positioned between Steps 6 and 7 of the algorithm. Another feature of the monotonicity test with respect to the unbounded case is dealt with in Sec. 4.6.

Let $X \in \mathbf{I}_\infty^m$, $f : X \to \mathbf{R}$, let $\dfrac{\partial f(x)}{\partial x_i}$ or $\partial f_i(x)$ be the i-th component of the gradient or the generalized gradient, resp., of f at x.

The generalized gradient of f at x was defined as

$$\partial f(x) = \text{conv }\{\lim\nolimits_{n \to \infty} f'(x_n) : x_n \to x, x_n \notin S \cup \Omega\} \qquad (4.13)$$

if at least one limit did exist, cf. Sec. 2.7. Again, conv denotes the convex hull, $f'(x_n)$ the gradient of f at x_n, Ω the set of points in some neighborhood of x at which f is not differentiable, and S any set of Lebesgue measure 0.

Let ∂f or f' exist in X and let $F_i : \overline{\mathbf{I}}(\overline{X}) \to \overline{\mathbf{I}}$ be an inclusion function of $\dfrac{\partial f}{\partial x_i}$ or of ∂f_i in the sense that

$$\dfrac{\partial f(x)}{\partial x_i} \in F_i(Y) \text{ or } \partial f_i(x) \in F_i(Y)$$

for any $x \in Y \cap \mathbf{R}^m$, $Y \in \bar{\mathbf{I}}(\overline{X})$. We set $Y_i = [a_i, b_i] \in \bar{\mathbf{I}}$ and $\overline{X}_i = [c_i, d_i] \in \bar{\mathbf{I}}$. Further, let $Y(i/s)$ for $s \in Y_i$ denote that box which arises from Y by replacing Y_i with s (or more precisely, with $[s, s]$). Then the *monotonicity test*, modified to operate on the unbounded case, consists of the following two parts (and it is primarily destined to handle the boxes $Y = V_1, V_2$ which arise by bisection in Step 6 of 1):

TEST 1 *Test for strictly monotone increasing: For some $i = 1, \ldots, m$, if $0 < \mathrm{lb}F_i(Y)$ then*

(i) *if $c_i < a_i$ then discard (Y, y) from the list,*

(ii) *if $c_i = a_i \in \mathbf{R}$ then replace (Y, y) with the pair (Y', y') where $Y' = Y(i/a_i)$ and $y' = \mathrm{lb}F(Y')$,*

(iii) *if $a_i = -\infty$ then terminate Alg. 1 (since $f^+ = -\infty$ such that problem (4.5) has no solution).*

TEST 2 *Test for strictly monotone decreasing: For some $i = 1, \ldots, m$, if $\mathrm{ub}F_i(Y) < 0$ then*

(i) *if $b_i < d_i$ then discard (Y, y) from the list,*

(ii) *if $b_i = d_i \in \mathbf{R}$ then replace (Y, y) with the pair (Y', y') where $Y' = Y(i/b_i)$ and $y' = \mathrm{lb}F(Y')$,*

(iii) *if $b_i = -\infty$ then terminate Alg. 1 (since $f^+ = -\infty$ such that problem (4.5) has no solution).*

Remarks. (1) It is a consequence of a termination by (iii) that only one global minimizer of X^+ is found. But, this is sufficient to guarantee the unsolvability of (4.5), cf. Theorem 3.

(2) It is favorable to admit $\pm\infty$ as values of the limits in the definition of the generalized gradient, (4.13). This is shown in the following example where even the bounded case is improved.

Example. Let the semicircle
$$f(x) = (1 - x^2)^{1/2}$$
be defined on $X = [-1, 1]$. It is obvious that $X^* = \{-1, 1\}$ and that $f^* = 0$. Standard methods of global optimization have difficulties in obtaining this result, since $f'(x) \to \mp\infty$ as $x \to \pm 1$.

We applied Alg. 1 with monotonicity test to this problem, using the extended generalized gradient as mentioned in Rem. (2). We took
$$F(Y) = (1 - Y^2)^{1/2} \text{ and } F'(Y) = -Y(1 - Y^2)^{1/2}$$
as inclusion functions for f and ∂f, resp., and got the exact result after 3 iterations. $F'(Y)$ has been computed via an infinite interval arithmetic as described in the next section. The 3 iterations were the following: List L_1 contained X only, L_2 contained $[-1, 0]$ and $[0, 1]$, and finally, L_3 contained $Z_1 = [-1, -1/2]$, $Z_2 = [-\frac{1}{2}, 0]$, $Z_3 = -Z_2$, $Z_4 = -Z_1$. Now, Z_1 and Z_4 were replaced with the exact minimizers, ± 1, by the monotonicity test, and Z_2 and Z_4 were discarded by the midpoint test.

4.5 Arithmetic in $\mathbf{I}_\infty$

This section develops a calculus of how to practically find the inclusions which are necessary for Alg. 1 to run. In contrast to the former sections, we no longer focus on topological properties (convergence), but we are now interested in a practicable arithmetic. The simplest way to get the inclusions for Alg. 1 is, first, to develop a calculus in $\mathbf{I}_\infty$ and to construct inclusion functions with respect to $\mathbf{I}_\infty^m$, and second, to extend these inclusion functions to $\overline{\mathbf{I}}^m$ by means of (4.3) and (4.4). That is, we admit only non-wasteful inclusion functions which, after all, is the natural way to deal with these problems.

The arithmetic in $\mathbf{I}_\infty$ is one with minimum requirements but adjusted for our purpose. As opposed to Kahan's (1968) or Laveuve's

(1975) arithmetic the values $\pm\infty$ are not accepted as points of intervals but they are used as boundaries of intervals. For instance, $\infty \notin [0, \infty) \in \mathbf{I}_\infty$. The transition to the case where $\pm\infty$ are points of intervals is done via the compactification, like $\infty \in [0, \infty] \in \bar{\mathbf{I}}$. Thus, if $A, B \in \mathbf{I}_\infty$ and if the product AB is to be defined, we need not take care of cases such as $0 \in A, \infty \in B$ which would involve defining inclusions for 0∞. This would inflate the arithmetic too much. We can therefore just set $0[0, \infty) = \{xy : x \in \mathbf{R}, y \in \mathbf{R}, x = 0, 0 \leq y < \infty\} = 0$, in contrast to other infinite arithmetics.

Let $A, B \in \mathbf{I}_\infty$. We expect our arithmetic in $\mathbf{I}_\infty$ to satisfy

$$A * B = \{a * b : a \in A, b \in B\} \tag{4.14}$$

if $*$ stands for $+, -$, and $\cdot$, and A/B to be the *smallest interval* of $\mathbf{I}_\infty$ or to be the *union of the two smallest intervals* of $\mathbf{I}_\infty$ such that

$$A/B \supseteq \{a/b : a \in A, b \in B, b \neq 0\}. \tag{4.15}$$

The case $A/0$ is excluded. (As usual, we write a instead of $[a, a]$ for brevity.)

For example, $1/[1, \infty) = [0, 1]$, but $\{1/b : b \in [1, \infty)\} = (0, 1] \notin \mathbf{I}_\infty$. Or, $1/[-1, 1] = (-\infty, -1] \cup [1, \infty)$. This means that $\mathbf{I}_\infty$ is not closed with respect to division. We do not worry about that and split up the union that occurs into two intervals of $\mathbf{I}_\infty$ and process the two intervals separately as long as necessary. For example,

$$Y := 1/[-1, 1] + [0, 2]/[1, 2]$$

may be evaluated in the following manner: Owing to our previous example and since $[0, 2]/[1, 2] = [0, 2]$, we split up the addition,

$$\begin{aligned} Y &= 1/[-1, 1] + [0, 2] = ((-\infty, -1] + [0, 2]) \cup ([1, \infty) + [0, 2]) \\ &= (-\infty, 1] + [1, \infty) = \mathbf{R} \in \mathbf{I}_\infty. \end{aligned}$$

In order to establish formulas for (4.14) and (4.15) which are usable, we introduce the following notation which avoids distinctions according to compact and non-compact intervals: Let $a, b \in \overline{\mathbf{R}}, a \leq b$.

We set
$$<a,b> := \begin{cases} [a,b] & \text{if } a,b \in \mathbf{R}, \\ (-\infty, b] & \text{if } a = -\infty, b \in \mathbf{R}, \\ [a, \infty) & \text{if } a \in \mathbf{R}, b = \infty, \\ (-\infty, \infty) & \text{if } a = -\infty,\ b = \infty. \end{cases}$$

Let now $<a,b>, <c,d> \in \mathbf{I}_\infty$. Then (4.14) and (4.15) are equivalent to
$$\left. \begin{aligned} <a,b> + <c,d> &= <a+c, b+d>, \\ <a,b> - <c,d> &= <a-d, b-c>, \\ <a,b>\!<c,d> &= <\min(ac, ad, bc, bd), \\ & \qquad \max(ac, ad, bc, bd)>, \end{aligned} \right\} \quad (4.16)$$

$$\left. \begin{aligned} & 1/<c,d> = [1/d, 1/c] \text{ if } 0 \notin <c,d>, \\ & 1/<0,d> = [1/d, \infty) \text{ if } d \neq 0, \\ & 1/<c,0> = (-\infty, 1/c] \text{ if } c \neq 0, \\ & 1/<c,d> = (-\infty, 1/c] \cup [1/d, \infty) \text{ if } c < 0 < d, \\ & <a,b>/<c,d> = <a,b>\cdot(1/<c,d>) \\ & \qquad \text{if } <c,d> \neq 0. \end{aligned} \right\} \quad (4.17)$$

Formulas (4.16) and (4.17) are only then well defined when an arithmetic for the boundaries a, b, c, d is given for the case that they are not finite. This is the usual one:
$$\mp\infty \mp \infty = \mp\infty,\ a \mp \infty = \mp\infty + a = \mp\infty \text{ if } a \in \mathbf{R},$$
$$0(\mp\infty) = (\mp\infty)0 = 0,\ +(\mp\infty) = \mp\infty,\ -(\mp\infty) = \pm\infty,$$
$$a(\mp\infty) = (\mp\infty)a = (\text{sgn}(a))(\mp\infty) \text{ if } a \neq 0,\ a \in \overline{\mathbf{R}},$$
$$a/\infty = 0 \text{ if } a \in \mathbf{R}.$$

(Expressions such as $0/0$, $a/0$, $\infty - \infty$, etc., do not occur in (4.16) and (4.17) and need not be defined.)

Example 1. $(-\infty, 1] + [2, \infty) = (-\infty, \infty) = \mathbf{R}.$

Example 2.
$[0,1](-\infty, 0] = <\min(0, 0, -\infty, 0), \max(\ldots)> = (-\infty, 0].$

If $f : D \to \mathbf{R}$, $D \subseteq \mathbf{R}^k$ is a function pre-declared in the programming language used (like sin, cos, etc.) and if $Y \in \mathbf{I}_\infty^k$ then the *natural interval extension* of f to Y denoted by $f(Y)$ is defined as the smallest interval of $\mathbf{I}_\infty$ such that

$$f(Y) \supseteq \square f(Y \cap D). \tag{4.18}$$

Practically, this definition causes no difficulties at all since the monotonicity domains of the functions usually pre-declared are well known such that the range of f over $Y \cap D$ can be determined easily. Thus, if $<a, b> \in \mathbf{I}_\infty$,

$$\begin{aligned}
\sin <a, b> &= \square \sin <a, b>, \\
\cos <a, b> &= \square \cos <a, b>, \\
\exp(<a, b>) &= <\exp(a), \exp(b)>, \text{ when } \exp(-\infty) := 0, \\
\ln(<a, b>) &= \square \ln(<a, b> \cap (0, \infty)), \\
\mathrm{sqr}(<a, b>) &= \square \mathrm{sqr}(<a, b>), \\
\mathrm{sqrt}(<a, b>) &= \square \mathrm{sqrt}(<a, b> \cap [0, \infty)).
\end{aligned}$$

etc.

Therefore, a natural interval extension of any programmable function f over $Y \in \mathbf{I}_\infty(X)$, $X \in \mathbf{I}_\infty^m$, can be defined recursively via (4.14), (4.15) and (4.18), in the same way as a function value $f(x)$ is defined recursively via the basic functions (arithmetic operations, pre-declared functions) - for example, by means of a computer code. The recursive representation of programmable functions is treated in detail by McCormick (1983), Rall (1981), and others. See also Sec. 2.6.

Example 3. If $f(x) = 1/(|\sin x| + |\cos x|)$ and if $Y = \mathbf{R}$ then the natural interval extension of f to $\mathbf{R}$ is $f(\mathbf{R}) = 1/(|\sin \mathbf{R}| + |\cos \mathbf{R}|) = 1/([0, 1] + [0, 1]) = 1/[0, 2] = [1/2, \infty)$. This result certainly depends on the representation chosen for $f(x)$, cf. Sec. 2.6.

Why is it necessary to admit boxes Y in (4.18) which are not necessarily contained in the domain of f? The reason is that, at each step of a recursive evaluation of a natural interval extension, an overestimation of the range is possible such that the domain of the function of the next recursive step can be exceeded. Such a superfluous overestimate is prevented in a natural way by the intersection $Y \cap D$.

As we already mentioned before it is one of the most fascinating research areas of interval mathematics to find, given any function $f : x \to \mathbf{R}$, $x \in \mathbf{I}^m$, an inclusion function F of f as good as possible; see Sec. 2.6. The same holds for functions $f : X \to \mathbf{R}, X \in \mathbf{I}_\infty^m$. In general, reasonable inclusion functions can be obtained by rearranging the given function expression of $f(x)$ and then by taking the natural interval extension of the rearrangement. The theory of such rearrangements is beyond our scope, but we emphasize that sometimes such rearrangements are necessary in order to meet the bounded box assumption of Theorem 2, cf. also the Example and Remark 2 both following Theorem 2.

4.6 Realization on the Computer

Up to now there is no generally widespread programming language in which an infinite interval arithmetic is incorporated. This is, however, no real problem for a programmer. In our case, one has to be aware that both kinds of infinite intervals we deal with must be approximated by appropriate intervals on the computer.

A simple way which we used and which is consistent with the languages mentioned above is the following: Intervals of the form $[a, L]$ and $[-L, b]$ where L is the largest real number representable on the computer under consideration are chosen to approximate both kinds of infinite intervals. These intervals get a special treatment.

In order to be more precise, let $\mathbf{R}_M$ be the set of machine representable real numbers. We assume that, for brevity, $-L$ is the smallest real number representable on the machine. Let

$$\overline{\mathbf{I}}_M = \{[a, b] : a, b \in \mathbf{R}_M, \ a \leq b\}$$

and

$$\mathbf{I}_M = \{[a, b] \in \overline{\mathbf{I}}_M : a \neq -L, \ b \neq L\}.$$

We call the intervals of $\mathbf{I}_M$ the *machine-finite* intervals and the intervals of $\overline{\mathbf{I}}_M \setminus \mathbf{I}_M$ the *machine-infinite* intervals. Both kinds of intervals constitute the machine intervals. The machine-finite intervals are interpreted as usual as sets

$$[a, b] = \{x : a \leq x \leq b\}.$$

The machine-infinite intervals are interpreted in two different ways depending on their purpose.

If we deal with arithmetic matters, construction of inclusions, etc., we want to have

$$[a, L] := \begin{cases} \mathbf{R} \text{ if } a = -L, \\ \{x : a \leq x < \infty\} \text{ if } -L < a \leq L, a \in \mathbf{R}_M, \end{cases} \quad (4.19)$$

$$[-L, b] := \{x : -\infty < x \leq b\} \text{ if } -L \leq b < L, b \in \mathbf{R}_M.$$

If we deal with topological properties, with the execution of Alg. 1, etc., we interpret the machine-infinite intervals in $\overline{\mathbf{I}}$ in the following manner:

$$[a, L] := \begin{cases} \overline{\mathbf{R}} \text{ if } a = -L, \\ \{x : a \leq x \leq \infty\} \text{ if } -L < a \leq L, a \in \mathbf{R}_M, \end{cases} \quad (4.20)$$

$$[-L, b] := \{x : -\infty \leq x \leq b\} \text{ if } -L \leq b < L, b \in \mathbf{R}_M.$$

The situation seems involved but is not. It even has the great practical advantage that, when transmitting the inclusions $F(Y)$ to $F(\overline{Y})$, the approximating machine intervals *need not be changed*. The step from $F(Y)$ to $F(\overline{Y})$ is just the compactification step connecting the arithmetic construction with the topological treatment by Alg. 1 such that the final interpretation of the solution set of Alg. 1 is done via (4.20).

An arithmetic $\tilde{*}$ in $\overline{\mathbf{I}}_M$ where $*$ stands for $+, -, \cdot,$ and $/$ is defined as follows: Let $A, B \in \overline{\mathbf{I}}_M$ then $A \tilde{*} B$ is the smallest interval of $\overline{\mathbf{I}}_M$ or the union of the two smallest intervals of $\overline{\mathbf{I}}_M$ such that

$$A \tilde{*} B \supseteq A * B. \quad (4.21)$$

In case of division, $B = 0$ is excluded. The interpretation of the machine-infinite intervals is done via (4.19). The requirements (4.21) can be fulfilled easily. See Kulisch-Miranker (1981) for a general theory of such requirements. It would be boring to give the precise instructions of how to implement (4.21) on the computer; rather, we give some examples,

$$[2,\ L] \tilde{+} [-L,\ L] = [-L,\ L],$$
$$[-2,\ L] \tilde{+} [L,\ L] = [L-2,\ L],$$
$$[2,\ L] \tilde{\cdot} [-L,\ -2] = [-L,\ -4],$$
$$1 \tilde{/} [-1,\ 1] = [-L,\ -1] \cup [1,\ L].$$

The approximation of the interval values for the functions pre-declared by intervals of $\bar{\mathbf{I}}_M$ is done analogously; for instance,

$$\ln[-1,\ L] = [-L,\ L],$$
$$\sin[1,\ L] = [-1,\ 1],$$

etc.

If, now, Alg. 1 runs on a machine then, after the computation is terminated, either the information $f^+ = -\infty$ (due to the monotonicity test) will be delivered or a machine interval $A \supseteq [\tilde{y}_n,\ f_n]$ and machine boxes $W_{ni} \supseteq Z_{ni}, i = 1, \ldots, l_n$, will be the output data. Here n is the final iteration index and l_n is the length of the list $\mathbf{L}_n$. In general, we get inclusions A and W_{ni} instead of $[\tilde{y}_n,\ f_n]$ and Z_{ni} because of the common outward rounding when a machine interval arithmetic is used. Owing to the execution of Alg. 1, we have $f^+ \in [\tilde{y}_n,\ f_n]$ and $X^+ \subseteq W := \bigcup_{i=1}^{l_n} W_{ni}$.

Using (4.20) and Theorem 3, the output data of the computation with Alg. 1 have to be interpreted as follows in order to get the required solution of problem (4.5):

1. If $f^+ = -\infty$ then f is unbounded from below and (4.5) has no solution.

2. If A is machine-finite and $W_{ni} \in \mathbf{I}_M^m, i = 1, \ldots, l_n$ (regular case), then problem (4.5) has a solution, f^* and X^*, and

$$f^* \in A,\ X^* \subseteq W.$$

3. If A is machine-infinite or if $W_{ni} \in \bar{\mathbf{I}}_M^m \setminus \mathbf{I}_M^m$ for at least one $i \in \{1, \ldots, l_n\}$ then a decision has not been possible whether a

solution of (4.5) does exist or not. However, *if* a solution exists, f^* and X^*, then

$$f^* \in A \cap \mathbf{R} \text{ and } X^* \subseteq W \cap \mathbf{R}^m.$$

Considering 1., one notices that the result $f^+ = -\infty$ is sharp, although the points from $-\infty$ to $-L$ cannot be distinguished on the computer. This is due to the monotonicity test which asserts that, if f is strictly monotonically increasing over the machine interval $[-L, a]$ representing $[-\infty, a]$, then, clearly, $f^+ = -\infty$, and no tolerance occurs.

4.7 Numerical Results

The following examples, taken from Ratschek-Voller (1988), were computed on an Apple IIe microcomputer. The programming language was PASCAL-SC.

Example 1. We consider the six hump camel back function, $f(x) = 4x_1^2 - 2.1x_1^4 + x_1^6/3 + x_1x_2 - 4x_2^2 + 4x_2^4$ for $x \in \mathbf{R}^2$. The inclusion function F as described in the Example after Theorem 2 (Sec. 4.3) was used for larger boxes Y and the meanvalue form (cf. Sec. 2.7) for F for the boxes Y which were machine-finite with $w(Y) \leq 1$. When the starting box was $\overline{\mathbf{R}}^2$ resp. $[-L, L]^2$ we needed 213 iterations of Alg. 1 with the monotonicity test (which is about 426 interval function evaluations of $F(Y)$) in order to get the intended absolute accuracy of 10^{-6} for the solution,

$$f^* \in -1.03162\ 84535\ 8 + [0, 5]10^{-11},$$
$$X^* \subseteq W_1 \cup W_2,$$

where
$$W_1 = [-8.98426\ 8, -8.98414\ 8]10^{-2} \times [7.12655\ 78, 7.12656\ 98]10^{-1},$$
$$W_2 = -W_1.$$

W_1 as well as W_2 contain exactly one global minimizer each (Moore-Ratschek (1987)).

By contrast, when Alg. 1 (also with monotonicity test) was applied to the same function with nearly the same inclusion function, but with the bounded box $[-2.5, 2.5]^2$, then 163 iterations (about 326 evaluations of $F(Y)$) were required in order to get a comparable result.

Example 2. P. Wolfe's (1975) function modified by Zowe (1985) is defined as

$$f(x) = \begin{cases} 5(9x_1^2 + 16x_2^2)^{1/2} & \text{if } x_1 \geq |x_2|, \\ 9x_1 + 16|x_2| & \text{if } 0 < x_1 < |x_2|, \\ 9x_1 + 16|x_2| - x_1^9 & \text{if } x_1 \leq 0, \end{cases}$$

where $x \in \mathbf{R}^2$. The only global minimizer of f with respect to $\mathbf{R}^2$ is $x^* = (-1, 0)$ and $f^* = -8$. The function is convex, and f fails to be differentiable only on the ray $x_1 \leq 0, x_2 = 0$. Zowe demonstrates that steepest descent with exact line search generates points convergent to the nonminimizer $(0, 0)$ when the starting point is chosen anywhere in the region $x_1 > |x_2| > (9/16)^2 |x_1|$.

Let $D_1 = \{x \in \mathbf{R}^2 : x_1 \geq |x_2|\}, D_2 = \{x \in \mathbf{R}^2 : 0 < x_1 < |x_2|\}$, and $D_3 = \{x \in \mathbf{R}^2 : x_1 \leq 0\}$. Let the functions $F_i : \mathbf{I}_\infty^2 \to \mathbf{I}, i = 1, 2, 3$, be defined as

$$\begin{aligned} F_1(Y) &:= 5(9Y_1^2 + 16Y_2^2)^{1/2}, \\ F_2(Y) &:= 9Y_1 + 16|Y_2|, \\ F_3(Y) &:= 9Y_1 + 16|Y_2| - Y_1^9. \end{aligned}$$

We extend f from $X = \mathbf{R}^2$ to $\overline{X} = \overline{\mathbf{R}}^2$ by means of (4.1) and consider the following inclusion function F of f on $\overline{X}$: Let $Y \in \mathbf{I}_\infty(X), w(\overline{Y}) \geq 1/2$. Then

$$F(Y) := \begin{cases} F_1(Y) & \text{if } Y \subseteq D_1, \\ F_2(Y) & \text{if } Y \subseteq D_2, \\ F_3(Y) & \text{if } Y \subseteq D_3, \\ F_i(Y) \cup F_j(Y) & \text{if } Y \subseteq D_i \cup D_j, Y \not\subseteq D_i, Y \not\subseteq D_j (i, j = 1, 2, 3), \\ \bigcup_{i=1}^3 F_i(Y) & \text{if } Y \not\subseteq D_j \cup D_k \text{ for } j, k = 1, 2, 3, \end{cases}$$

$$F(\overline{Y}) := \overline{F(Y)}.$$

For $Y \in \mathbf{I}(X), w(Y) < 1/2$, we use the meanvalue form of f on Y as inclusion function value $F(Y)$, cf. Sec. 2.7, where inclusions of the generalized gradient instead of inclusions of the derivative are taken if no derivative is available.

When Alg. 1 (with monotonicity test) was applied to F and $\overline{X} = \overline{\mathbf{R}}^2$, then 104 iterations (about 208 evaluations of $F(Y)$) were needed in order to determine the solutions x^* and f^* within an absolute accuracy of $2 \cdot 10^{-6}$.

Example 3. Let $X = [-1, 1] \times [-1, 1] \times [0, \infty) \subseteq \mathbf{R}^3$ and $f : X \to \mathbf{R}$ be defined by

$$f(x) = (1 - x_1^2)^{1/2} \cos x_3 + (1 - x_2^2)^{1/2}/(1 + x_3^2) + 2x_3 e^{-x_1}.$$

There exist 4 global minimizers of f in X having the coordinates $x_1 = \pm 1, x_2 = \pm 1, x_3 = 0$. The objective function, f, is differentiable in the interior of X, continuous - but not Lipschitz - on X. However, f is generalized differentiable (when infinite values are admitted) on the edge of X which contains the four minimizers, such that the monotonicity test can be performed and the meanvalue form of f can be constructed when inclusions of the generalized gradients are used instead of inclusions of non-existing derivatives.

As inclusion function $F(Y)$ of f on $\overline{X}$ (extended by means of (4.1)) the plain natural interval extension of f on Y was used if $Y \in \mathbf{I}_\infty(X)$ or if $w(Y) \geq 1$ and further, $F(\overline{Y}) = \overline{F(Y)}$, in these cases. For $Y \in \mathbf{I}(X)$ and $w(Y) < 1$ the meanvalue form of f on Y was taken for $F(Y)$.

Alg. 1 (with monotonicity test) needed 31 iterations (which makes about 62 evaluations of $F(Y)$) in order to achieve the intended absolute accuracy of about 10^{-6} for X^* and f^*.

Chapter 5

Constrained Optimization

5.1 Introduction

It is frequently the case that a mathematical model of a system contains limitations on the acceptable values of the parameters of the system induced by the particular configuration considered. This is, for example, often the case in engineering design, chemical equilibrium calculations, agricultural models, economic models and so on. The limitations or constraints may be nonlinear. This leads to the most general global optimization topic, namely the global nonlinear constrained optimization problem.

In this chapter this general problem will be dealt with. We keep in mind that it is not our aim to discuss all possible or "best" methods for solving this problem. Rather we want to give a sample of how several interval and non-interval methods fit together and can be combined to an effective and reliable algorithm.

The reader may also have a look at the algorithms of Sengupta (1981), Hansen-Sengupta (1980), (1983), Hansen-Walster (1987a,b). These algorithms are interesting and highly sophisticated alternatives which try to process as much information as possible.

An effective procedure for the global constrained optimization problem will consist of

(i) the basic algorithm,

(ii) the accelerating devices.

The basic algorithm is mainly responsible for getting the solution of the problem or, at least, an approximate solution when the computation is done on a computer. The accelerating devices aim to get the solution or its approximation as fast as possible.

As in the unconstrained case, the midpoint test is included in the basic algorithm since this test is responsible for the fact that the solution can be obtained as sharply as desired. The midpoint test also speeds up the computation since it protects the algorithm from processing superfluous areas. Therefore the midpoint test should be applied as early as possible in the computation. Since the midpoint test can only be applied if feasible points are known we focus on methods which create primarily a feasible point. Therefore, the algorithm we discuss will consist of the following:

(a) an initial search for a feasible point,

(b) starting from a feasible point a search is initiated for a local minimum or for feasible points with lower values of the objective function,

(c) the basic algorithm which mainly consists of the exhaustion principle (areas are exhausted that cannot contain the solution) which needs no differentiability assumptions,

(d) if appropriate differentiability or generalized differentiability conditions are given then gradient methods or Newton techniques can be applied to discover areas where the function is monotone, or to localize local minimizers.

The chapter is organized in the following manner. The global constraint minimization problem is defined and the notation is given (Sec. 5.2). An algorithm for finding the feasible domain is given (Sec. 5.3). An exact solution cannot be found in some cases owing to the algorithmic approach. In order to overcome the related difficulties a relaxed algorithm is established (Sec. 5.4). The algorithm for finding a feasible domain is combined with the basic algorithm for the unconstrained case in order to have an always working reliable basic algorithm for the constrained case (Sec. 5.5). The constraint conditions are sometimes not given precisely. This leads to interval constraints (Sec. 5.6). The convergence properties of the basic algorithms are developed in Sec. 5.7. Finally, the accelerating devices are discussed depending on the differentiability conditions of the functions of the problem (Sec. 5.8 - 5.11).

5.2 Problem Statement

For convenience we repeat some of the definitions of Sec. 1.2. Let $X \in \mathbf{I}^m$ be the domain where the optimization problem is discussed. Let f, g_i, h_j be real-valued functions defined on X, where $i = 1, \ldots, k$ and $j = k + 1, \ldots, r$. The *global constrained optimization problem* (better: *minimization problem*) consists of

(i) finding the minimum value, f^*, of f over X when the arguments x satisfy the restrictions

$$g_i(x) \leq 0, \quad i = 1, \ldots, k \tag{5.1}$$

and

$$h_j(X) = 0, \quad j = k+1, \ldots, r, \tag{5.2}$$

and of

(ii) finding the set X^* of arguments x^* satisfying (5.1), (5.2) and $f(x^*) = f^*$. Instead of (5.1) and (5.2) we sometimes prefer the vector notation, $g(x) \leq 0$ and $h(x) = 0$, where the constraint functions g_i and h_j are seen as components of the vector-valued functions $g : X \to \mathbf{R}^k$ and $h : X \to \mathbf{R}^{r-k}$.

f is called the *objective function* of the problem; the conditions (5.1) and (5.2) are called the *inequality constraints* and the *equality constraints* of the problem. Accordingly, g_i and h_j are the *inequality* resp. *equality constraint functions*. The set D of all $x \in X$ satisfying (5.1) and (5.2) is called the *feasible* set of the problem. The points of D are called *feasible* points. Points which are not feasible are called *infeasible*. An area which only consists of feasible [infeasible] points is called *feasible* [*infeasible*]. The value f^* is called the *global minimum* and the points $x^* \in X^*$ are called the *global minimizers* or the *global minimum points*.

The *global constrained minimization problem* is usually written in the following form,

$$\min_{x \in X} f(x) \text{ subject to } g(x) \leq 0, \quad h(x) = 0. \tag{5.3}$$

A point $x \in X$ is called a *local minimizer* or a *local minimum point* of problem (5.3) if x is feasible and if there exists an $\epsilon > 0$ such that

$$f(x) \leq f(y) \text{ for any feasible } y \text{ with } \| y - x \| < \epsilon$$

where any norm $\| \ \|$ of $\mathbf{R}^m$ can be chosen. If x is a local minimizer of (5.3) then $f(x)$ is called a *local minimum* of (5.3).

5.3 Constraints and Exhaustion Principle

The global constrained optimization problem, (5.3), involves both inequality and equality constraints.

If g is a set of linear inequalities and if there are no equality constraints then it is well-known that D is a convex polygon. If in addition the objective function f is linear then a linear programming problem is obtained whose solution is normally found at a vertex of the polygon. We do not discuss this special problem.

In general the set D may take an arbitrary non-convex shape when the constraints g and h are non-linear which means that D can not in general be calculated exactly in a finite number of steps.

There are many methods that can be used to solve the constraint problem. Four favorite approaches are

(i) Kuhn-Tucker and related methods,

(ii) penalty methods,

(iii) exhaustion methods,

Constraints and Exhaustion Principle

(iv) statistical methods.

The Kuhn-Tucker methods aim to solve the Kuhn-Tucker conditions, cf. (1.4) and (1.5), in order to obtain the local minimizers which include the global minimizers.

Penalty methods reduce the constrained case to a sequence of unconstrained problems.

Exhaustion methods check the whole area under consideration piecewise for a certain property, for example, whether the points are feasible or not. The area can thus be reduced to a smaller one by eliminating the pieces with the unwanted property, for example, the infeasible pieces.

Statistical methods are very effective, but there is only a certain percentage of reliability on the result. The percentage may tend to 100 as the number of steps of the computation grows.

The choice of the method will depend on the dimension of the problem, the differentiability and stability conditions, the shape of D and other data, and on the personal preference. Finally, the chosen method will be supported by additional techniques such as local information, accelerating devices, etc.

We use the exhausting principle as the underlying approach for solving the constrained problem. The reasons are that the exhausting principle makes it possible to check the whole box systematically, it needs no differentiability assumptions and it is easy to combine with accelerating techniques if available.

The exhaustion principle is applied twice: First, it is used to discard infeasible areas. This is discussed in this section. Second, it is used to discard areas that cannot contain the global minimizer. This is done in Sec. 5.5. The main tool that makes it possible to use the exhaustion principle in a very effective manner is interval arithmetic.

It is assumed that g and h have inclusion functions G and H, that is, $\Box g(Y) \subseteq G(Y)$ and $\Box h(Y) \subseteq H(Y)$ for each $Y \in \mathbf{I}(X)$.

The aim of this section is to determine D, the feasible set. Although D will not be needed explicitly in this monograph we develop the computation of D thoroughly in order that the reader should become familiar with the exhaustion principle. As we can see in Sec. 5.7, Alg. 1 which is introduced below has D as solution set under very mild conditions, that is, g and h are continuous, and $w(G(Y)) \to 0$ as well as $w(H(Y)) \to 0$ when $w(Y) \to 0$. Practically, Alg. 1 will stop after a finite number of calculations, and D cannot be reached. In this case, however, an approximating set U is produced which has the following properties:

(1) $D \subseteq U$,

(2) $U = \bigcup_{i=1}^{l} Z_i$ where $Z_i \in \mathbf{I}(X)$ for some index l,

(3) $\mathrm{lb}G(Z_i) \leq 0, i = 1, \ldots, l$,

(4) $0 \in H(Z_i), i = 1, \ldots, l$,

(5) $U \setminus D$ is smaller than a given tolerance.

The boxes Z_i, cf. (2), are stored in a list by the algorithm. Thus, l is the length of this list.

The set U is restricted to a finite union of boxes by (2). Hence $U \neq D$ in general.

Since the set U contains the feasible set it follows that the set $V = X \setminus U$, the complement of the set U in X, is included in the infeasible set $X \setminus D$. This means that the set V is an infeasible set which may be excluded from further considerations in the global constrained minimization problem.

Suppose now that a feasible inclusion U has been computed. Then the set $U \setminus D$ is an infeasible region included in U, the estimate of the feasible region. The set $U \setminus D$ is an overestimate region.

This region may in general not be reduced to the empty set without further assumptions for G and H. The reason for this is that independently of the subdivision technique the inclusion functions G and H may also introduce an uncertainty by overestimation since $\Box g(Y) \subseteq G(Y)$, $\Box h(Y) \subseteq H(Y)$ and $0 < \mathrm{ub}G(Y)$ or $H(Y) \neq 0$ can occur even if Y is feasible.

The region of uncertainty clearly leads to a termination criterion: For this let

$$Z^* = \bigcup \{Z_j \in U | \mathrm{lb}G_i(Z_j) \leq 0 < \mathrm{ub}G_i(Z_j) \text{ for some } i = 1, \ldots, k \\ \text{or } 0 \in H_i(Z_j) \neq 0 \text{ for some } i = k+1, \ldots, r\}.$$

Constraints and Exhaustion Principle

Then if $w(Z^*) < \epsilon$ for some predefined ϵ the computation shall stop. Such a termination criterion makes sense if the assumptions for G and H mentioned above hold.

If the assumptions mentioned are not fulfilled, the properties (1) to (4) still hold.

The algorithm is initialized with the box X and G, H. It then subdivides X and tests whether the inequality and equality constraints hold in each box created by the subdivision. Hence the name "exhausting principle".

When a box Y has been created by subdivision the inequality constraint inclusions $G_i(Y)$, $i = 1, 2, \ldots, k$, and the equality constraint inclusions $H_i(Y)$, $i = k+1, \ldots, r$, are computed. The feasibility or the infeasibility of boxes Y can only be determined via the inclusions G and H:

(i) If $G(Y) \leq 0, H(Y) = 0$ then Y is feasible,

(ii) if $G_i(Y) > 0$ for some i then Y is infeasible,

(iii) if $0 \notin H_i(Y)$ for some i then Y is infeasible.

In all the other cases, a decision cannot yet be made. Thus, we call a box Y *indeterminate* (subject to G and H) if (i) to (iii) do not hold.

Alg. 1 stores a box Y if it satisfies (i) since Y is feasible. Together with Alg. 2, cf. Sec. 5.5, it can be checked whether Y contains any global minimizers. Box Y is discarded if it satisfies (ii) or (iii) since Y is infeasible; in this case Y cannot contain any minimizers such that we need not store Y for later use. If Y is indeterminate, Y will be subdivided such that a decision may be possible for subregions of Y.

If $G_i(Y) \leq 0$ for some i then $g_i(x) \leq 0$ for $x \in Y$, that is, the i-th constraint is satisfied in Y. Thus, for any subbox $Y_o \subseteq Y$, $G_i(Y_o)$ need not be computed again, since $g_i(x) \leq 0$ for $x \in Y_o$. The same holds for H_i. Such an information is handled via flags. Thus, a flag vector

$$R = (R_1, \ldots, R_r)$$

is attached to each box under consideration where the components R_i take the values 0 or 1. The vector R is used as follows:

$R_i = 1$ indicates that $g_i(x) \le 0$, if $i = 1,\ldots,k$, or that $h_i(x) = 0$, if $i = k+1,\ldots,r$, for any $x \in Y$.

$R_i = 0$ indicates that, up to the current state of the computation, $g_i(x) \le 0$ resp. $h_i(x) = 0$ for any x has not been verified.

In later subdivisions of Y, the flags $R_i = 1$ are transferred to the subboxes, and recalculations of $G_i(Y_o)$ or $H_i(Y_o)$ need only take place if $R_i = 0$.

The algorithm generates and manipulates a list L of pairs (Y, R) where R is the flag vector which describes the current feasibility situation for Y. $R = (1,\ldots,1)$ means that Y is feasible, and Y is indeterminate if $R_i = 0$ for at least one index i. If boxes Y are recognized as infeasible they are discarded so there is no need for a flag to denote such boxes.

The algorithm works by always subdividing the first indeterminate box on the list (having a flag vector where $R_i = 0$, for some i) if it exists. This box is the oldest indeterminate box of the list. After the subdivision there are three possibilities for each of the two new boxes. If a new box is feasible it is added to the end of the list with a flag vector $R = (1,\ldots,1)$. If the box is indeterminate then it is entered onto the list immediately after all the indeterminate boxes on the list. Infeasible boxes are discarded.

The following algorithm contains a relaxation parameter $\epsilon_1 \ge 0$. First we admit no freedom and set $\epsilon_1 = 0$. In this case the algorithm works as explained. The impact of values $\epsilon_1 > 0$ is explained in the next section.

ALGORITHM 1 *The computation of a set including the feasible region.*

1. *Set $Y := X$.*
2. *Set $R = (R_1,\ldots,R_r) := (0,\ldots,0)$.*
3. *Initialize list $L := ((Y, R))$.*
4. *Choose a coordinate direction ν parallel to which $Y_1 \times \ldots \times Y_m$ has an edge of maximum length, i.e. $\nu \in \{i : w(Y) = w(Y_i)\}$.*

Constraints and Exhaustion Principle

5. *Bisect Y normal to direction ν, getting boxes V_1, V_2 such that $Y = V_1 \cup V_2$.*

6. *Remove (Y, R) from the list L while retaining R. {Remark: $R_i = 0$ for some i .}*

7. *For $j = 1, 2$*
   (a) *Set $R^j = (R_1^j, \ldots, R_r^j) := R$.*
   (b) *For $i = 1, 2, \ldots, k$ if $R_i = 0$ then*
      i. *Calculate $G_i(V_j)$.*
      ii. *If $G_i(V_j) > 0$ then go to (e).*
      iii. *If $G_i(V_j) \leq 0$ then set $R_i^j := 1$.*
   (c) *For $i = k+1, k+2, \ldots, r$ if $R_i = 0$ then*
      i. *Calculate $H_i(V_j)$.*
      ii. *If $0 \notin H_i(V_j)$ then go to (e).*
      iii. *If $H_i(V_j) \subseteq [-\epsilon_1, \epsilon_1]$ then set $R_i^j := 1$.*
   (d) *Enter (V_j, R^j) onto the list after all the items having flag vectors $R \neq (1, \ldots, 1)$ and in front of the items having flag vectors $R = (1, \ldots, 1)$.*
   (e) *end (of j-loop).*

8. *If list is empty then terminate with output:*
   - $D \neq \emptyset$.

9. *Denote the first pair of the list by (Y, R).*

10. *If the termination criteria hold then go to 12.*

11. *Go to 4.*

12. *There are two possibilities for the boxes $Z_1, \ldots, Z_l$ on the final list:*

   A: If no flag vector R on the final list is equal to $(1, \ldots, 1)$ then the output is:
   - *It was not possible to decide whether D (the feasible domain) is empty or not.*
   - *If it is the case that $D \neq \emptyset$ then $D \subseteq \bigcup_{i=1}^{l} Z_i$.*

B: If at least one flag vector R is $(1,\ldots,1)$ then the output is:

- $D \neq \emptyset$
- $D \subseteq \bigcup_{i=1}^{l} Z_i$.

13. End.

At termination the list contains some boxes that are feasible and some boxes that are indeterminate.

This algorithm has been presented in order to demonstrate the principle of the basic subdivision and exhaustion method for solving a set of constraints having only an inclusion function. It is, however, only computationally meaningful in combination with other algorithms. The separate description of Alg. 1 is also necessary for its convergence theory; see Sec. 5.7. The exhaustion principle of Alg. 1 is also applied if nonlinear equations are being solved that have curves, surfaces or other geometrical figures as a solution set, cf. Neumaier (1988).

5.4 Trouble with Constraints

Relaxing the constraints

If the optimization problem one has to solve involves equality constraints then no matter how long the computation lasts the algorithm may fail to produce even one feasible box Y. That is, in such cases

$$H(Y) = 0$$

must be verified which is mathematically as well as computationally almost impossible. For example, let $X = [0,1] \times [0,1]$, and let the feasible points $x = (x_1, x_2) \in X$ be characterized by $x_1 = x_2$. Thus, the feasible domain D is the diagonal of X. Alg. 1 will never produce a feasible box. This example indicates a typical situation and is no exemption. If the algorithm runs on a computer then, additionally, rounding errors will also prevent obtaining a result like $H(Y) = 0$. There is no general way out of this disaster. Nevertheless, a very important and far-reaching exception of this embarrassing situation is described in the second part of this section.

These circumstances are usually ignored by books on optimization. The general strategy is that, if a point x satisfies the constraints within a certain tolerance, x is accepted as a feasible point. Alg. 1 does not ignore these facts, and no uncertain points are accepted as feasible points, but the user gets the message that the existence of feasible points cannot be proven *via the computation*. If the user, however, knows that $D \neq \emptyset$ (may be, due to theoretical investigations, or topological or analytical considerations, or due to practical reasons such as physical observations) then he is correct when using the result that D is included in the union of the remaining boxes Z_i, cf. Step 12 of Alg. 1.

If the user does not intend to check whether a feasible point exists or not, but if he only requires some properties of the feasible points then the user may, assuming $D \neq \emptyset$, relax the equality constraints and replace them with

$$-\epsilon_1 \leq h_i(x) \leq \epsilon_1 \text{ for } x \in X, i = k+1, \ldots, r,$$

for some $\epsilon_1 > 0$. Thus the equality constraints are replaced by pairs of inequality constraints. The advantage of their use is that the algorithm may sooner find a "feasible" area which is of importance for the combined algorithms described in the sequel since the midpoint test can then be applied. The inequality constraints may also be relaxed in such cases. This could be done automatically by the program if no feasible box has been produced after a reasonable amount of computation has been done. This indicates that the feasible domain is probably of a lower dimension like a surface, hyperplane, etc., or of small size. Relaxations such as

$$g_i(x) - \epsilon_1 \leq 0, \quad i = 1, \ldots, k,$$

are reasonable.

If the algorithm is applied to the relaxed problem then the existence of feasible points can result. These feasible points are related to the relaxed and not to the original problem. However, the resulting unions, $\bigcup_{i=1}^{l} Z_i$, are inclusions of the set of feasible points of the original problem too.

Let us finally return to the termination criterion of Alg. 1 mentioned in the last section. The criterion is based upon the assumption that the area of Z^*, which is the set of indeterminate points, tends to 0 as the computation continues. For conditions implying this assumption, see Sec. 5.7. There are rare cases where this assumption is not satisfied. To avoid such cases one can incorporate a side termination criterion which always causes a stop of the computation, for example:

Terminate when $w(Y) < \epsilon$ for all indeterminate boxes Y of a current list.

Certainly, the printouts and results of Alg. 1 remain valid under this criterion.

One way out

As mentioned before there is one way to overcome the difficulties arising by lower-dimensional feasible areas, that is the application of Moore's (1977) *test for the existence of solutions of equations*. Hansen-Walster (1987a) were the first to suggest that this test should be applied to constrained optimization. The assumptions under which this test works are that

(i) the constraint functions which cause problems are of $\mathbf{C}^1$,

(ii) there are not too many troublesome constraints.

Let Y be an indeterminate box. In order to prove that Y contains a feasible point we proceed as follows: Without restriction of the generality, let

$$\text{lb}G_i(Y) \leq 0 < \text{ub}G_i(Y) \text{ for } i = 1,\ldots,\kappa(\leq k),$$
$$0 \in H_i(Y) \neq 0 \text{ for } i = k+1,\ldots,k+\rho(\leq r),$$
$$s := \kappa + \rho.$$

The remaining constraints shall not be violated,

$$G_i(Y) \leq 0 \quad \text{for } i = \kappa+1,\ldots,k,$$
$$H_i(Y) = 0 \quad \text{for } i = k+\rho+1,\ldots,r.$$

We need the following assumptions:

(i) $g_1, \ldots, g_\kappa, h_{k+1}, \ldots, h_{k+\rho}$ are $\mathbf{C}^1$-functions,

(ii) $s \leq m$.

We choose s components of Y - for simplicity the first s - and build a subbox of Y,
$$\tilde{Y} := Y_1 \times \ldots \times Y_s.$$
If $s < m$ then $\tilde{Y}$ is part of the edge of Y. Let further $\tilde{c} := \text{mid } \tilde{Y}$, $c_i := \text{mid } Y_i$ for $i = s+1, \ldots, r$, and
$$\tilde{x} := (x_1, \ldots, x_s).$$

The assignments
$$\phi_i(\tilde{x}) := \begin{cases} g_i(\tilde{x}, c_{s+1}, \ldots, c_m), & i = 1, \ldots, \kappa, \\ h_{i+\kappa-k}(\tilde{x}, c_{s+1}, \ldots, c_m), & i = k+1, \ldots, k+\rho \end{cases}$$

for $\tilde{x} \in \tilde{Y}$ define an s-dimensional vector-valued function
$$\phi = (\phi_1, \ldots, \phi_s)^T : \tilde{Y} \to \mathbf{R}^s.$$

Let $J_\phi(\tilde{x})$ denote the Jacobian matrix of ϕ at $\tilde{x} \in \tilde{Y}$, and $J(\tilde{Y}) \in \mathbf{I}^{s \times s}$ be an interval matrix that contains the matrices $J_\phi(\tilde{x})$ for all $\tilde{x} \in \tilde{Y}$. For instance, $J(\tilde{Y})$ could be the natural interval extension of $J_\phi(\tilde{x})$ to $\tilde{Y}$, or $J(\tilde{Y})$ could be constructed using the methods of Sec. 2.8 such that $J(\tilde{Y})$ has a smaller number of interval entries.

We consider the equation
$$\phi(\tilde{c}) + J(\tilde{Y})(\tilde{x} - \tilde{c}) = 0$$

with respect to the variable $\tilde{x} \in \tilde{Y}$. Let S be the solution set of this equation as defined in Sec. 2.9. A theorem of Bao-Rokne (1987), which is a generalization of Moore's (1977) existence theorem, says that if $S \subseteq Y$ then ϕ has a solution $\tilde{x}_0$ in Y, that is $\phi(\tilde{x}_0) = 0$. In practice it does not matter that neither S nor $\tilde{x}_0$ is known. In order to verify $S \subseteq Y$ it is sufficient, for instance, to apply one iteration of the interval Newton method to ϕ in Y, that is Step 2 of the algorithm as presented in Sec. 2.9. (This step can be executed, for instance, with the relaxation procedure of Sec. 2.10, with the elimination procedure of Sec. 2.10, or even with one iteration of the complete Hansen-Greenberg realization.) The iteration will yield a superset Z of S. If now $Z \subseteq Y$ then $S \subseteq Y$ and the assumption is verified. In this case we know that a feasible point $x_0 := (\tilde{x}_0, c_{s+1}, \ldots, c_m)^T$ exists in Y, that is, $g(x_0) \leq 0$ and $h(x_0) = 0$. Therefore, D is not empty.

Owing to the construction of ϕ we know that this feasible point x_0 has $g_1, \ldots, g_\kappa$ as active constraints and $g_{\kappa+1}, \ldots, g_k$ as inactive constraints. If therefore the interval Newton iteration results in $Z \cap Y = \emptyset$ which means that ϕ has no solution in Y (cf. Sec. 2.9), then one cannot conclude that Y has no feasible point. One can only conclude that there is no feasible point in Y having $g_1, \ldots, g_\kappa$ as active constraints.

The knowledge that Y has a feasible point and that D is not empty has no direct influence on Alg. 1, but it is of importance in the sequel when the existence of feasible points is necessary for the application of the midpoint test and other accelerating devices. Therefore one will not apply the existence test in connection with Alg. 1 too early since the smaller the box Y is the likelier the existence test yields a positive result. One will further choose a box for the test which has a flag vector R with as few zero components as possible.

5.5 The Basic Algorithm

In this section the basic algorithm for solving the constrained problem (5.3) is developed. It is a minimum algorithm and it is recommended that it is combined with steps that accelerate the speed of the computation. The algorithm is, however, described separately in order to make the influence of interval arithmetic more transparent, in order to give the reader the possibility of combining the basic algorithm with his special choice of accelerating devices - a few of which are described in the following sections - and, finally, since the convergence properties, cf. Theorem 2, are addressed exactly to this algorithm due to our notation of a basic algorithm in Sec. 5.1.

The following Alg. 2 combines Alg. 1 (for obtaining the feasible domain) with Alg. 3 of Ch. 3 (Hansen's basic algorithm for solving the global unconstrained case). Since Alg. 1 as well as Alg. 3 of Ch. 3

The Basic Algorithm

use the exhaustion principle, Alg. 2 does as well. Alg. 2 is, however, not constructed by appending Alg. 3 of Ch. 3 to Alg. 1. Instead the steps of the two contributing algorithms are merged together in order to avoid repeated processing of one and the same box. The main features of Alg. 2 are:

- It uses subdivision (exhaustion principle) for the localization of the global minimizers and global minimum (Step 6 of Alg. 2).

- It discards both feasible and infeasible boxes using the midpoint test (Step 11 of Alg. 2).

- It uses interval evaluation of the constraints to discard infeasible boxes (one constraint violated) or to find feasible boxes (all constraints satisfied). It only recalculates constraints previously undecided (Step 8 of Alg. 2).

The aim of Alg. 2 is to determine X^*, the set of global minimizers, and f^*, the global minimum of the constraint problem (5.3). As we can see in Sec. 5.7, Alg. 2 has X^* and f^* as solution set under very mild conditions, that is, f, g, and h are continuous, the inclusion functions F, G, and H of f, g, and h resp. satisfy the usual contraction conditions, $w(F(Y)) \to 0$, $w(G(Y)) \to 0$ and $w(H(Y)) \to 0$ as $w(Y) \to 0$, and further, some topological conditions. Practically, Alg. 2 will stop after a finite number of operations, and X^*, f^* cannot be reached. In this case, approximating data $(U, \tilde{y}, \tilde{f})$ are produced which have the following properties:

(1) $X^* \subseteq U$,

(2) $U = \bigcup_{i=1}^{l} Z_i$ where $Z_i \in \mathbf{I}(X)$ for some index l,

(3) $U \setminus X^*$ is smaller than a given tolerance,

(4) $\tilde{y} \leq f^* \leq \tilde{f}$ if a solution exists,

(5) The difference $\tilde{f} - \tilde{y}$ (error estimate) is smaller than a given tolerance, if $D \neq \emptyset$, where D denotes the feasible region.

The boxes Z_i, cf. (2), are just the boxes which are in the list at the termination. Thus, l is the length of the final list. The value $\tilde{y}$ arises from Step 10 of Alg. 2. It is the only lower bound v_j, cf. Step 8(b), guaranteed to be an underestimate of f^*. The value $\tilde{f}$ arises from Steps 2 and 8(f). It is the smallest function value of a feasible point which has been computed. If $D = \emptyset$ then $\tilde{f} = \infty$.

If the assumptions mentioned above do not hold, properties (1), (2) and (4) are nevertheless valid. Also (3) and (5) remain frequently valid.

Points (3) and (5) can be used for termination criteria, as extensively discussed in the unconstrained case. Possible termination criteria are thus:

(i) Terminate if $\tilde{f} - \tilde{y} < \epsilon$.

(ii) Terminate if $\sum_{i=1}^{l} w(Z_i) < \epsilon$ where Z_i are the boxes of the final list and l its length.

(iii) Terminate if $w(Z_i) < \epsilon$ where Z_i are as in (ii).

Criterion (i) causes a stop if the convergence assumptions are fulfilled. Criterion (ii) causes a stop if the convergence assumptions are satisfied *and* the m-dimensional Lebesgue measure of X^* is smaller than ϵ. Criterion (iii) always works. At numerical computation it can happen that (i) and (ii) do not work because of an accumulation of rounding errors. In such cases ϵ has to be chosen larger or (iii) may be used.

The following basic algorithm for solving the global constraint minimization problem (5.3) has as input data the dimension of the problem, m, the domain X of the problem, inclusion functions F, G, and H of f, g, and h, and a feasible point $\tilde{x}$ if available. Finally a *relaxation parameter* $\epsilon_1 \geq 0$ is provided. If $\epsilon_1 = 0$ then problem (5.3) is solved. If $\epsilon_1 > 0$ then the relaxed problem is solved which arises from (5.3) by replacing the equality constraints $h_j(x) = 0$, $x \in X$, by inequality constraints $h_j(x) \in [-\epsilon_1, \epsilon_1]$, $x \in X$, for $j = k+1, \ldots, r$, cf. Sec. 5.4.

ALGORITHM 2 *The basic algorithm for solving the global constrained optimization problem.*

The Basic Algorithm

1. Set $Y := X$.
2. If a feasible point $\tilde{x}$ is given set $\tilde{f} := \mathrm{ub}F(\tilde{x})$ else set $\tilde{f} := \infty$.
3. Set $R = (R_1, \ldots, R_r) := (0, \ldots, 0)$; set $y := \mathrm{lb}F(Y)$.
4. Initialize list $\mathbf{L} := ((Y, y, R))$.
5. Choose a coordinate direction ν parallel to which $Y_1 \times \cdots \times Y_m$ has an edge of maximum length, i.e. $\nu \in \{i : w(Y) = w(Y_i)\}$.
6. Bisect Y normal to direction ν, getting boxes V_1, V_2 such that $Y = V_1 \cup V_2$.
7. Remove (Y, y, R) from the list $\mathbf{L}$ while retaining R.
8. For $j = 1, 2$
    (a) Set $R^j = (R_1^j, \ldots, R_r^j) := R$.
    (b) Calculate $F(V_j)$ and set $v_j = \mathrm{lb}F(V_j)$.
    (c) If $\tilde{f} < v_j$ then go to (h).
    (d) For $i = 1, 2, \ldots, k$ if $R_i = 0$ then
        i. Calculate $G_i(V_j)$.
        ii. If $G_i(V_j) > 0$ then go to (h).
        iii. If $G_i(V_j) \leq 0$ then set $R_i^j := 1$.
    (e) For $i = k+1, k+2, \ldots, r$ if $R_i = 0$ then
        i. Calculate $H_i(V_j)$.
        ii. If $0 \notin H_i(V_j)$ then go to (h).
        iii. If $H_i(V_j) \subseteq [-\epsilon_1, \epsilon_1]$ then set $R_i^j := 1$.
    (f) If $R^j = (1, \ldots, 1)$ then set $\tilde{f} := \min(\tilde{f}, \mathrm{ub}F(c_j))$ where $c_j = \mathrm{mid}\, V_j$.
    (g) Enter (V_j, v_j, R^j) onto end of list.
    (h) End (of j-loop).
9. If list $\mathbf{L}$ is empty then terminate with output:
    - $D = \emptyset$ (no feasible point).
10. Choose a triple $(\tilde{Y}, \tilde{y}, \tilde{R})$ from the list which satisfies $\tilde{y} \leq z$ for all triples (Z, z, R).

11. (Midpoint test). *Discard all triples* (Z, z, R) *on the list that satisfy* $\tilde{f} < z$.

12. *Denote the first pair of the list by* (Y, y, R).

13. *If termination criteria hold go to 15.*

14. *Go to 5.*

15. *There are two possibilities for the boxes* $Z_1, \ldots, Z_l$ *on the final list:*

   A: *If no flag vector* R *on the final list is equal to* $(1, \ldots, 1)$ *then the output is:*
   - *It was not possible to decide whether* D *(the feasible domain) is empty or not.*
   - *If it is the case that* $D \neq \emptyset$ *then*
   $$X^* \subseteq D \subseteq \bigcup_{i=1}^{l} Z_i,$$
   $$\tilde{y} \leq f^*.$$

   B: *If at least one flag vector* R *is* $(1, \ldots, 1)$ *then the output is:*
   - $D \neq \emptyset$
   - $X^* \subseteq \bigcup_{i=1}^{l} Z_i$
   - $\tilde{y} \leq f^* \leq \tilde{f}.$

16. *End.*

If the number of indeterminate boxes does not shrink during the continued course of the algorithm and if the difference $\tilde{f} - \tilde{y}$ does not decrease, it is likely that some region is left for which it cannot be decided whether it is feasible or not in spite of the repeatedly executed bisections. These phenomena are mainly due to the geometric shape of D and they will almost always occur if equality constraints are involved in the optimization problem.

By assigning a positive value to the relaxation parameter ϵ_1 results may be obtained that might be useful for the user. In this case one must again be aware that the solutions refer to the relaxed and not to the originally posed problem. Nevertheless such solutions are frequently more helpful than results which are too far from the correct solution to be realistic.

Therefore we recommend to incorporate the *existence test* described in Sec. 5.4 whenever possible. It is reasonable to submit the box $\tilde{Y}$ to the test or another box Z having a low value $z = \mathrm{lb}F(Z)$. If the test confirms the existence of a feasible point in a box Z then the value

$$\mathrm{ub}F(Z)$$

is a *reliable upper bound* of f^* and appropriate for updating $\tilde{f}$ by

$$\tilde{f} := \min(\tilde{f}, F(Z)).$$

The test and the subsequent updating are best inserted between Steps 10 and 11 of the algorithm. The midpoint test (Step 11) following the test can then already use the hopefully improved value $\tilde{f}$.

If the existence test is applied periodically, convergence can be obtained even though the conditions for convergence, cf. Sec. 5.7, do not hold.

5.6 Optimization Problems with Inexact Data

Frequently, the functions f, g, and h which occur in problem (5.3) cannot be given exactly. This may be the case when they originate from observations, measurements, physical experiments, or even from previous numerical computations. For example, if the earth acceleration, light velocity, etc. are part of such functions, the function values will not be real numbers but tolerances of real numbers. This means that the function values are intervals, a situation which may arise for all functions involved in (5.3).

In this section we show how one can solve such ill-posed problems and subordinate them to our general approach for solving (5.3). See also Hansen (1984).

Our inexactly posed problem is the following: Solve (5.3), that is, solve

$$\min_{x \in X} f(x) \text{ subject to } g(x) \leq 0, h(x) = 0$$

where f, g, and h are *not exactly known*. The assumption is that instead of f, g and h tolerance functions

$$\phi : X \to \mathbf{I}, \psi : X \to \mathbf{I}^k \text{ and } \chi : X \to \mathbf{I}^{r-k}$$

are known where

$$f(x) \in \phi(x), \quad g(x) \in \psi(x) \text{ and } h(x) \in \chi(x)$$

for all $x \in X$.

Certainly, owing to the ill-posed problem statement, an exact solution cannot be expected even with an "ideal" algorithm. Since, however, ϕ, ψ and χ act in some manner as inclusion functions for f, g, and h one can apply the logical relationships between functions and inclusion functions as in the previous sections. Thus, if $x \in X$ and if

$$\text{ub}\psi(x) \leq 0, \chi(x) = 0$$

then x is feasible (with respect to problem (5.3), naturally). If

$$\begin{aligned} &\text{lb}\psi_i(x) > 0 &&\text{for some } i = 1, \ldots, k \text{ or} \\ &0 \notin \chi_i(x) &&\text{for some } i = k+1, \ldots, r \end{aligned}$$

then x is infeasible. If x satisfies none of these conditions then a decision as to whether x is feasible or not is not possible, and x remains indeterminate.

In order to make the ill-posed problem accessible to Alg. 1 and Alg. 2, we need inclusion functions F, G, and H for ϕ, ψ, and χ. This means that

$$\square\phi(Y) \subseteq F(Y), \quad \square\psi(Y) \subseteq G(Y), \quad \square\chi(Y) \subseteq H(Y)$$

for any $Y \in \mathbf{I}(X)$. This implies that F, G, and H are also inclusion functions for f, g, and h and the methods and algorithms of this chapter can be applied directly.

That is, if Alg. 1 is applied to X, F, G, and H, then D or inclusions U of D will be produced. Also the results are possible that no feasible point exists or that a decision whether a feasible point exists cannot be given and an inclusion for the eventually existing feasible set D is given. Analogously, if Alg. 2 is applied.

Since $\phi(x) \subseteq F(x)$, etc., for $x \in X$, which means that, in general, $w(F(X)) > 0$ even though $w(Y) = 0$ for $Y = x$, the convergence conditions $w(F(Y)) \to 0$ as $w(Y)$ etc. will not be satisfied, cf. Sec. 5.7. Thus, the feasible set D or f^* and X^* cannot be approximated as well as one would have wished.

5.7 Convergence Properties

In this section the mathematical background for Alg. 1 and Alg. 2 is given. We supply sufficient conditions under which the solution set of Alg. 1 is D, the set of feasible points of the global constraint optimization problem, and under which the solution set of Alg. 2 is the set of global minimizers, X^*, and the global minimum, f^*, of problem (5.3). The accelerating devices of the following sections have no influence on the solution sets of the algorithms.

The main assumptions we need are the contraction properties of the inclusion functions in the same manner as in the unconstrained case. If again F, G and H are the inclusion functions of f, g, and h, as defined in (5.3), we consider the following assumptions,

$$w(F(Y)) \to 0 \text{ as } w(Y) \to 0 \text{ for } Y \in \mathbf{I}(X), \tag{5.4}$$

$$w(G(Y)) \to 0, \ w(H(Y)) \to 0 \text{ as } w(Y) \to 0 \text{ for } Y \in \mathbf{I}(X). \tag{5.5}$$

(5.4) implies the continuity of f, and (5.5) implies the continuity of g and h.

Notation like $U_n \to D$ means $d(U_n, D) \to 0$ where U_n and D are non-empty compact sets and d is the Hausdorff-metric of compact sets; see Sec. 3.2.

Let us start with Alg. 1. We use the notation of Sec. 5.3 but add subscripts where necessary. Let $Z_{n1}, \ldots, Z_{nl_n}$ be the boxes which are in L_n, that is, the list at the n-th iteration. Let

$$U_n = \bigcup_{i=1}^{l_n} Z_{ni}.$$

We then have the following theorem.

THEOREM 1 *Let Alg. 1 be applied to the box X, the relaxation parameter $\epsilon_1 = 0$ and the inclusion functions G and H of g and h. If the contraction assumptions (5.5) hold, then the sequence (U_n) forms a nested sequence and $\bigcap_{n=1}^{\infty} U_n = D$ which means that $U_n \to D$ if $D \neq \emptyset$.*

Proof. It is obvious that (U_n) is nested. In order to get a contradiction for the remaining assertion we assume that the point $x \in X$ is infeasible but that $x \in U_n$ for all n. Let W_n be one box of the list L_n that contains x. Since the bisection of a box is done normal to the direction of a largest edge, one gets $w(W_n) \to 0$ as $n \to \infty$, cf. Lemma 1 of Ch. 3. This implies $W_n \to x$ as $n \to \infty$. Since $g_i(x) > 0$ for some $i = 1, \ldots, k$ or $h_i(x) \neq 0$ for some $i = k+1, \ldots, r$, we have $G_i(W_n) \to g_i(x)$ resp. $H_i(W_n) \to h_i(x)$ as $n \to \infty$ due to (5.5). This means that for some n, $\mathrm{lb} G_i(W_n) > 0$ resp. $0 \notin H_i(W_n)$ such that the algorithm recognizes W_n as infeasible and deletes W_n. Thus W_n cannot be a box of list L_n. This gives the contradiction. □

Let us now investigate how Alg. 2 works. Again let $Z_{ni}, \ldots, Z_{nl}$ be the boxes of list L_n and $U_n = \bigcup_{i=1}^{l_n} Z_{ni}$ where l_n is the length of L_n; see Sec. 5.5. Let $\tilde{y}_n$ be the current value of $\tilde{y}$ in list L_n, and let f_n be the current value of $\tilde{f}$ in list L_n.

The convergence properties of Alg. 2 depend on the possibility of applying the midpoint test as often as desired in order to exhaust all points which are not global minimizers. This leads to the following assumption which is mainly of a topological character:

$$\left. \begin{array}{l} \text{There exists a sequence of points } x_n \text{ lying in} \\ \text{the interior of the feasible domain } D \text{ and} \\ \text{converging to some global minimizer } x^* \in X^*. \end{array} \right\} \text{(T)}$$

It is certainly difficult to verify (T) *a priori* since neither D nor X^* is known. The disadvantage of such assumptions is, however,

Convergence Properties

shared by convergence proofs of other, also non-interval, algorithms for global constraint problems. A sufficient condition for (T) is, for example, that $D \neq \emptyset$ is connected and that the closure of the interior of D contains D. Another sufficient condition is that at least one global minimizer lies in the interior of D, etc.

The following theorem discusses the convergence properties of Alg. 2.

THEOREM 2 *Let Alg. 2 be applied to the box X, the relaxation parameter $\epsilon_1 = 0$ and the inclusion functions $F, G,$ and H of $f, g,$ and h, resp. Let the contraction assumptions (5.4), (5.5) and condition (T) be satisfied. Then the sequence (U_n) is nested and $\bigcap_{n=1}^{\infty} U_n = D$ which means that $U_n \to D$. Furthermore, $\tilde{y}_n \to f^*$ with $\tilde{y}_n \leq f^*$ and $f_n \searrow f^*$ as $n \to \infty$.*

Proof. It follows directly from the construction of Alg. 2 that (U_n) is nested and that $U_n \supseteq D$ for all n. It is further obvious that, for any n,

$$\tilde{y}_n \leq f^*, \tag{5.6}$$

$$f^* \leq f_n, \tag{5.7}$$

$$f_{n+1} \leq f_n. \tag{5.8}$$

Since the bisection of the boxes is executed in the same way as in the unconstrained case we have

$$w(Y_n) \to 0 \text{ as } n \to \infty;$$

see proof of Lemma 1 of Ch. 3. Furthermore since $w(Y_n) \geq w(Z_{ni})$ for $i = 1, \ldots, l_n$, we get

$$w(Z_{ni}) \to 0 \text{ as } n \to \infty \tag{5.9}$$

independent of i.

We will first show that

$$\tilde{y}_n \to f^*. \tag{5.10}$$

Now, $\tilde{y}_n = \text{lb}F(\tilde{Y}_n) \leq z_{ni}$ for all $i = 1, \ldots, l_n$. Since $w(\tilde{Y}_n) \to 0$ due to (5.9), it follows from (5.4) that $w(F(\tilde{Y}_n)) \to 0$. Since $f^* \in$

$F(\tilde{Y}_n)$ we get (5.10). We justify the use of $f^* \in F(\tilde{Y}_n)$ later. If this formula was not true then we would have $\tilde{y}_n \to f^\circ < f^*$ for some $f^\circ \in R$ because of (5.6). This would imply $F(\tilde{Y}_n) \to f^\circ$ which would finally mean $\mathrm{ub}F(\tilde{Y}_n) < f^*$, that is, $\tilde{Y}_n$ would be infeasible for sufficiently large n. Since $w(G(\tilde{Y}_n)) \to 0$ as well as $w(H(\tilde{Y}_n)) \to 0$ and since f is continuous which is implied by (5.4), $\tilde{Y}_n$ would have been discarded for large n by the exhaustion of infeasible domains. This proves (5.10). Note that for the proof of (5.10) the midpoint test and thus assumption (T) are not needed.

We show next that
$$f_n \searrow f^*.$$
Owing to (5.8) it suffices to show that
$$f_n \to f^*. \tag{5.11}$$

Let (x_n) be a sequence converging to some $x^* \in X^*$ as is specified by (T). We can find a subsequence $(f_{k_n})_{n=1}^\infty$ of $(f_n)_{n=1}^\infty$ that satisfies
$$f_{k_n} \leq f(x_n) \text{ for each } n \tag{5.12}$$
as follows:

Given x_n, there are two possibilities: Either

(i) there exists a box Z_{k_n} of list L_{k_n} such that $x_n \in Z_{k_n}$ and $\mathrm{ub}G(Z_{k_n}) \leq 0, H(Z_{k_n}) = 0$ or

(ii) there does not exist such a box as described in (i).

In case (i) we have $f_{k_n} \leq \mathrm{lb}F(Z_{k_n}) \leq f(x_n)$ owing to the updating procedure of $\tilde{f}$ in Alg. 2. In case (ii), we consider (5.5) and (5.9) which implies that a list L_{k_n} exists such that no box on the list contains x_n. Since x_n is feasible the boxes of the former lists that contained x_n must have been discarded by the midpoint test, i.e. $f_{k_n} < f(x_n)$, owing to (5.8). Thus, in both cases we have shown that
$$f_{k_n} \leq f(x_n).$$
Since $x_n \to x^*$ we get $f(x_n) \to f^*$ owing to the continuity of f. This shows, together with $f^* \leq f_{k_n} \leq f(x_n)$ and (5.8), the assertion (5.11).

It remains to show that $\bigcap_{n=1}^{\infty} U_n = X^*$. In order to get a contradiction let us assume that, for some $x \notin X^*$, we have $x \in U_n$ for all n. This means that for any n a box Z_n belonging to a list L_n exists with $x \in Z_n$. Since $w(Z_n) \to 0$ by (5.5), we get $Z_n \to x$ and further, $F(Z_n) \to f(x)$. Since $f_n \searrow f^*$ and $f(x) > f^*$, the boxes Z_n are thus discarded for large n by the midpoint test. This gives the contradiction. $\square$

Remarks. (1) The convergence property

$$\tilde{y}_n \nearrow f^*$$

does not depend on the midpoint test and hence on (T). Only the contraction assumptions (5.4) and (5.5) are necessary for the proof.

(2) If the optimization problem involves equality constraints then the feasible domain will in general be a hypersurface with an empty interior with respect to the m-dimensional topology. The condition (T) therefore does not apply. Accordingly the relaxation is an important means for reestablishing a reasonable topological situation in the form of (T). If relaxation is undesirable or not possible then the periodical application of the existence test will frequently force convergence even under unfavorable circumstances.

5.8 Accelerating Devices: Overview

The basic algorithm given in Sec. 5.5 may be modified in a variety of ways by the incorporation of acceleration devices as discussed in the introduction to this chapter. We distinguish two types of acceleration devices:

- preprocessing devices,
- internal devices.

The purpose of the preprocessing devices is to prepare a good starting position for the basic algorithm and they are applied to the input data before the basic algorithm starts. The internal devices are added within the iterations of the basic algorithm such that they are called up repeatedly.

The first preprocessing device is the search for an initial feasible point. If such a feasible point is known prior to the basic algorithm (Alg. 2) then the computation will be much more effective. The reason for this is that the initial minimum function value is then set to the value of f at the feasible point (or an upper bound of the objective function at the feasible point - see Step 2 of Alg. 2). This value is then used in the midpoint test to eliminate boxes that may not contain any global minimizers thus preventing superfluous processing. If such a value is not known then the midpoint test cannot be applied.

Finding a feasible point is in itself a nonlinear programming problem where numerical difficulties may be encountered if the set of constraints is ill-conditioned. For example, if the constraints are

$$g_1(x) = x_1 + \delta - x_2 \leq 0$$

and

$$g_2(x) = -x_1 + \delta + x_2 \leq 0$$

then there are no feasible points (x_1, x_2) if $\delta > 0$, one feasible point (x_1, x_2) if $\delta = 0$, and a continuum of feasible points if $\delta < 0$. If $|\delta|$ is small or zero then it might be difficult to decide whether a feasible point exists or not.

For this reason it is recommended that the initial search for a feasible point be terminated whenever

- a feasible point has been found *or*

- it has been shown that there are no feasible points *or*

- it has not been possible to decide whether a feasible point exists or not, but enough effort has been spent on this initial search.

The reason for terminating the process with an undetermined result is that although knowledge of an initial feasible point speeds up the computation it is not essential for the functioning (and correctness) of Alg. 2.

The precise statement of the search for a feasible point is:

Accelerating Devices: Overview

Given a box $X \in \mathbf{I}^m$ and constraint functions g, h as in problem (5.3), find a point $\tilde{x} \in X$ such that $g(\tilde{x}) \leq 0$ and $h(\tilde{x}) = 0$ or prove that such a point does exist in some subbox Y of X.

There are a number of techniques for computing such a point (if it exists). Some of these techniques which fit nicely into our interval arithmetic approach are discussed in the sequel depending on the differentiability of the objective function and the constraints. The value at the feasible point will be denoted by $\tilde{f}$ and it may be used directly in the midpoint test or may be used as a starting point for the devices mentioned in the next paragraph.

The second set of preprocessing devices consists of techniques for finding new feasible points $\tilde{x}$ with $f(\tilde{x}) < \tilde{f}$ in order to increase the effectiveness of the midpoint test. For this purpose we choose a few methods for finding a local minimizer for unconstrained optimization, as in Ch. 3, in the sequel. These methods typically determine a direction and a steplength from an initial point with the hope that the new point will provide a smaller function value. For the purpose of remaining in the feasible domain the methods are enhanced with a test for feasibility. If the new point is infeasible the stepsize is reduced in an iterative manner using a bisection procedure.

The internal devices include the procedures used to improve feasible points in order to enlarge the effectiveness of the midpoint test as in the preprocessing case. Tests for discarding subboxes directly (monotonicity test, interval Newton method) are counted as internal devices.

We further distinguish between

- local devices,

- global devices.

The local devices use point information and consist of classical non-interval procedures (search methods, gradient methods, Newton-like methods). They are used to improve the feasible points locally, that is, in a neighborhood of some feasible point already known The global devices use the whole information available in a current subbox and thus they are interval procedures. They are mainly used to discard subboxes directly or to determine all local and hence the global minimizers (interval Newton method). It is typical of the global devices that they can only be merged with the basic algorithm if the area to which they are applied is feasible.

Several devices were already treated in Chs. 2 and 3 such that our representation can be shortened in these cases.

5.9 Devices for Functions without Differentiability Properties

It is assumed here that both the objective function f and the constraint functions g, h are nonsmooth or not even continuous. This precludes the use of simple linearization techniques and all the more popular enhancements are therefore not available. If either f or g and h are nonsmooth one may combine appropriate devices from the following overview.

(A) Finding a feasible point

In the absence of smoothness of the constraint functions it is reasonable to focus on exhaustion techniques in connection with a branching – but no bounding – principle and the use of inclusion functions for the constraint functions.

We therefore propose a *preprocessing exhaustion algorithm* for finding a feasible point. It is applied to the box X and the constraint functions g and h of problem (5.3). This algorithm is positioned before the basic algorithm (Alg. 2). The preprocessing algorithm works in the following manner: By repeated bisection of the indeterminate boxes (these are the boxes which have not been computationally recognized as feasible or as infeasible) and the deletion of infeasible boxes the chances of finding a feasible point increase. In contrast to Alg. 1 we are not interested in having convergence to the feasible domain of the given problem; instead we wish to pick out just one feasible point as soon as possible.

Technically, this algorithm maintains a list of indeterminate boxes starting with a list of one box, the problem area X. With each box the flag vector R as introduced for Alg. 1 is attached. R keeps track of which constraints are satisfied. The computation is finished when a feasible point is found or when it does not make sense to continue the iterations, cf. the termination criteria discussed below.

In order to reach a feasible point as fast as possible we search for it in boxes where the chances are optimal for finding one. The chances are obviously best in boxes which satisfy a maximal number of constraints, and if there are several boxes with the same maximal number we take that box which violates the remaining constraints minimally. This is the branching idea which we will pursue. It can be realized in the following manner:

A counter κ keeping track of the number of satisfied constraints is maintained for each box. If several boxes have the same number of satisfied constraints then a measure due to Hansen-Sengupta (1983) will decide the priority and is computed as follows. If the current box is Y then for each i with $R_i = 0$ (meaning Y is indeterminate)

$$\alpha_i = \frac{\mathrm{ub} G_i(Y)}{w(G_i(Y))}$$

or

$$\beta_i = \frac{\mathrm{ub} H_i(Y)}{w(H_i(Y))}$$

is computed. The measure α for the box Y is now defined as the maximum of the computed values α_i and β_i. A box with the smallest α is selected among the boxes with the same maximum κ for further processing.

Each time a box is chosen from the list L for processing, a test is made to see if the midpoint is feasible. If so, the algorithm terminates with this point and the list of indeterminate boxes as output data.

Further, an input parameter ϵ_2 is provided. Its value may be chosen with the intention that boxes of width less than ϵ_2 are not to be subdivided further. The reason for this is that a box may contain points that satisfy κ constraints for some $\kappa < r$, but never satisfy further constraints, which means that the subdivisions should be stopped for that box if the width is too small. The box will be processed further in the main algorithm in any case such that no information is lost.

The algorithm also terminates if a prescribed number of iterations has been reached. The reason for this is that the main algorithm will continue the processing in a different manner and possibly report the case A in Step 15 of Alg. 2.

The algorithm will therefore be terminated if either

- a feasible point has been found *or*

- a limiting number of iterations has been reached *or*

- the boxes of the list have width smaller than some value ϵ_2, as discussed above.

The input parameter ϵ_1 has the same meaning as in Algs. 1 and 2 and one has to set $\epsilon_1 = 0$ if completely reliable results are required.

This results in the following algorithm.

ALGORITHM 3 *Preprocessing for getting a feasible point for the global constrained optimization algorithm.*

1. *Set $Y := X$.*

2. *Set $R = (R_1, \ldots, R_r) := (0, \ldots, 0)$, $\kappa = 0$, $\alpha = 0$.*

3. *Initialize list $L := ((Y, R, \kappa, \alpha))$.*

4. *Choose a coordinate direction ν parallel to which $Y_1 \times \cdots \times Y_m$ has an edge of maximum length, i.e. $\nu \in \{i : w(Y) = w(Y_i)\}$.*

5. *Bisect Y normal to direction ν getting boxes V_1, V_2 such that $Y = V_1 \cup V_2$.*

6. *Remove (Y, R, κ, α) from the list L while retaining R and κ.*

7. *For $j = 1, 2$*

   (a) *Set $R^j = (R_1^j, \ldots, R_r^j) := R$, $\kappa_j := \kappa$.*

   (b) *If $w(V_j) < \epsilon_2$ then set $\alpha_j = -1$ and $\kappa_j = 0$ and enter $(V_j, R, \kappa_j, \alpha_j)$ onto end of list and go to (g).*

   (c) *Set $\alpha := 0$.*

(d) For $i = 1, 2, \ldots, k$ if $R_i = 0$ then
  i. Calculate $G_i(V_j)$.
  ii. If $G_i(V_j) > 0$ then go to (g).
  iii. If $G_i(V_j) \leq 0$ then set $R_i^j := 1$ and $\kappa_j := \kappa_j + 1$.
  iv. If $R_i^j = 0$ then set $\alpha := \max(\alpha, \frac{\mathrm{ub}G_i(Y)}{w(G_i(Y))})$.

(e) For $i = k+1, k+2, \ldots, r$ if $R_i = 0$ then
  i. Calculate $H_i(V_j)$.
  ii. If $0 \notin H_i(V_j)$ then go to (g).
  iii. If $H_i(V_j) \subseteq [-\epsilon_1, \epsilon_1]$ then set $R_i^j := 1$ and $\kappa_j := \kappa_j + 1$.
  iv. If $R_i^j = 0$ then set $\alpha := \max(\alpha, \frac{\mathrm{ub}H_i(Y)}{w(H_i(Y))})$.

(f) Enter $(V_j, R^j, \kappa_j, \alpha)$ onto the list as the first element if $\kappa_j > \kappa$; otherwise enter onto list in order of decreasing κ. If several elements on resulting list have same κ then order these with respect to decreasing α.

(g) End (of j-loop).

8. If list L is empty then terminate with output:

   - $D = \emptyset$ (no feasible point exists).

9. Denote the first pair of the list by (Y, R, κ, α).

10. Set $c = \mathrm{mid}\, Y$.

11. If c is feasible then terminate with output:

    - c is feasible.

    or if $\alpha = -1$ then terminate with output:

    - No feasible point found and all boxes on list L have width less than ϵ_2.

    or if iteration limit exceeded then terminate with output:

    - No feasible point found and iteration limit exceeded.

12. Go to 4.

(B) Improving the feasible point

We propose here the same primitive search procedure as in the unconstrained case, cf. Sec. 3.12. The starting point is either the midpoint of the current (feasible) box Y or the feasible point produced by the preprocessing algorithm for finding a feasible point. Again, it does not matter if the current box Y is left by the search, but we are not allowed to leave the feasible domain $D \subseteq X$. In the case where the search in a certain direction leads outside of D one can repeat the search a few times in this direction with reduced steplengths.

As in the unconstrained case, subgradient or bundle methods etc. are available if f is locally Lipschitz in D.

(C) Further improvements

The only advice we can give is to invest as much work as possible in order to construct a good inclusion function.

If, however, f is locally Lipschitz the monotonicity test is applicable and the meanvalue form can be chosen as an inclusion function for F. The latter is also possible for g and h when these functions are locally Lipschitz.

5.10 Devices for C^1 Functions

It is now assumed that the objective function f or the constraints g and h are C^1 functions. This opens up the possibility of using a number of techniques based on linearization.

(A) Finding a feasible point

Since the constraints may be linearized it is possible to use an extension of the usual Newton method for finding zeros of a nonlinear equation to search for a feasible point. This extension was suggested by Robinson (1972) and Daniel (1973) and it results in the following method.

Devices for C^1 Functions

The Robinson-Daniel Method

1. Choose $x_0 \in X$.

2. For $n = 0, 1, 2, \ldots$

   (a) Choose x_{n+1} to minimize $\|y - x_n\|$ over the set S_n of points y satisfying
   $$g(x_n) + J_g(x_n)(y - x_n) \leq 0$$
   $$h(x_n) + J_h(x_n)(y - x_n) = 0.$$

   (b) If x_{n+1} is feasible then go to 3.

   (c) If other terminating conditions hold then go to 3.

3. End.

The norm used in the minimization procedure is the Euclidean norm. With this assumption a unique solution x_{n+1} exists since S_n is defined via linear equalities and linear inequalities, provided $S_n \neq \emptyset$. The finding of a feasible point is thus referred to the solving of a sequence of so-called *convex-simplex problems* (c.f. for instance Zangwill (1969)). Such problems are reduced to linear programming problems via the linearization of the convex objective function $\phi(y) = \|y - x_n\|$.

Another approach which is quite attractive because of its simplicity is the approach via a smooth *penalty function* where the objective function part is dropped, cf. Sec. 1.4. In the simplest case one computes one zero $\tilde{x}$ of the function

$$\Phi(x) = \sum_{i=1}^{k}(\max(g_i(x), 0))^2 + \sum_{i=k+1}^{r}(h_i(x))^2$$

using Newton's method or related methods. This zero $\tilde{x}$ is certainly a feasible point. If several attempts to get a feasible point with a non-interval Newton algorithm fail, Alg. 3 together with the existence test (cf. Sec. 5.4 and 5.5) has the best chance of finding a feasible point or to prove its existence in a subbox sufficiently small. In general, a failure of local methods to get a feasible point will be observed if equality constraints are involved in the optimization problem (5.3). In this case it will also be difficult to improve feasible points with local methods since these are based on linearizations which will lead out of the feasible domain. This makes the repeated application of the existence test necessary, which may prove the existence of improved feasible points.

Nevertheless, trouble as described in the previous paragraph can be avoided if the whole optimization problem (5.3) is reduced to the unconstrained optimization problem (3.1) by using the penalty model approach of Sec. 1.4.

(B) Improving a feasible point

As in the unconstrained case we propose gradient based methods for finding a lower function value. One only has to take care that the feasible domain D is not left when iteratively constructing the next point. Reducing the stepwidth is a simple means for staying within D. The Fletcher-Reeves algorithm, for instance, is then modified as follows:

The Fletcher-Reeves Method

1. *Given $x_0 \in D$ and set $p_0 := -f'(x_0)$.*

2. *For $n = 0, 1, 2, \ldots$*

    (a) *Let ρ_n be a solution of the problem $\min_{\rho \geq 0} f(x_n - \rho p_n)$.*

    (b) *Set $x_{n+1} := x_n + \rho_n p_n$.*

    (c) *If x_{n+1} is feasible then go to (e).*

    (d) *If x_{n+1} is infeasible then set $\rho_n := \rho_n/2$ and go to (b).*

    (e) *Set $p_{n+1} := -f'(x_{n+1}) + \beta_n p_n$ where $\beta_n = \dfrac{\|f'(x_{n+1}))\|^2}{\|f'(x_n)\|^2}$.*

    (f) *If $f'(x_{n+1})$ is small enough then go to 3.*

3. *End.*

Certainly, Step 2(d) should be executed only a few times.

(C) Further improvements

The meanvalue form is suggested for all kinds of inclusions needed whenever the box width is small. It is applied independently of the feasibility of the boxes since good inclusions are required in all cases.

The monotonicity test is applicable for feasible boxes only. There is no need to apply it to infeasible boxes since infeasible boxes are discarded anyway. Applying this test to indeterminate boxes can lead to wrong results since a strictly monotone f can take its global minimum when descending from a feasible part to an infeasible part.

5.11 Devices for C^2 Functions

First of all, the techniques for C^1 functions are available and they can be combined with the techniques of this section. Second, if equality constraints occur in the optimization problem (5.3), the best way to solve (5.3) is to apply the interval Newton method to the Fritz John or to the Kuhn-Tucker conditions. This approach is to be combined with the midpoint test in order to exclude solutions of these conditions which are not global minimizers. The interval Newton method will produce subboxes Y of X, the *union* of which contains the solutions. In order to guarantee the existence of solutions in the individual boxes the existence test of Sec. 5.4 and 5.5 may be applied to the boxes Y singly. Thus one obtains comparative values which are necessary for the midpoint test. The existence test needs no extra numerical effort since the data which are involved in the test are a by-product of the iterations of the Newton algorithm.

The first to apply the interval Newton method to the John conditions were Hansen-Walster (1987b). They report that a very effective procedure results, which is superior if the global minimizers are edge points of the feasible domain or near the edge.

If equality constraints are missing or if it is preferred to solve the problem (5.3) in a relaxed manner, cf. Sec. 5.3, the following devices are useful:

(A) Finding a feasible point

There are several possibilities for finding a feasible point:

If the constraint functions are not too involved then the solving of the Kuhn-Tucker or Fritz John conditions is promising, when the objective function f is set as identically zero, cf. Sec. 1.3.

For example, owing to the theory of Lagrangian multipliers, each feasible point solves the Fritz John conditions, that is, is a solution of the equations (1.2) where $f' \equiv 0$.

Newton-like methods may be used to solve these equations. Certainly they can fail. In this case we do not recommend the interval Newton method since the organizational effort is too high, and it would be better if interval Newton methods are used to solve the Fritz John conditions with the original objective function f and, not as we do here, with vanishing f.

If f, g and h are not too complicated one would also apply the local Newton method to the complete John conditions to get a local minimizer. Then one can skip the "improving the feasible point" part since nothing could be improved locally.

If one insists that a feasible point should be found while not wishing to use interval methods then one should consider Schnabel's (1982) paper where an interesting method for $\mathbf{C}^2$ functions is developed.

(B) Improving a feasible point

Typically, Newton's method for obtaining a local minimizer or one of its variants may be used. The advantage of these methods is that both a direction and a displacement are computed directly and may be used immediately without further calculations if the new point is feasible. A further advantage is the high rate of local convergence.

The Newton method *for improving a feasible point*

1. *Given x_0 (feasible).*

2. *For $n = 0, 1, 2, \ldots$*

    (a) *solve $f'(x_n) + f''(x_n)(x_{n+1} - x_n) = 0$ with respect to x_{n+1}.*

(b) If x_{n+1} is feasible go to (d).

(c) If x_{n+1} is not feasible then set $x_{n+1} := x_n + (x_{n+1} - x_n)/2$ and go to (b).

(d) End (of the n-loop).

If, for some n, a certain number of stepwidth reductions in (c) is not successful then the computation should be stopped. One may also stop the computation if it turns out that the x_n's tend to a local maximizer by checking the function values $f(x_n)$. Further if x_0 is not close to a local minimizer the method can encounter considerable difficulties.

(C) Further improvements

Best results have been obtained if – as in the unconstrained case – each iteration of the basic algorithm was combined with just one iteration of the interval Newton method when the current box was feasible. As mentioned in (B), this approach is slow if the global minimizers are near the edge of the feasible domain. In this case the approach via the Fritz John conditions is preferable.

5.12 Numerical Examples

Example 1. (Levy-Gomez (1985), p. 243, Problem 2). We consider the two-dimensional problem

$$minimize\ f(x) := 0.1(x_1^2 + x_2^2)$$

s. t.
$$g_1(x) := 2\sin(2\pi x_2) - \sin(4\pi x_1) \leq 0$$

over the domain $([-1, 1], [-1, 1])^T$. The problem has at least 24 local minima. The feasible domain is also not connected. There is exactly one global minimizer $x^* = (0, 0)^T$ with $f(x^*) = 0$.

The problem was solved using the basic algorithm Alg. 2 for the constrained problem. The inclusions for f and g_1 were computed using natural interval extensions.

The input data were:

$X = ([-1, 1], [-1, 1])^T$ as starting domain,

$\epsilon = E - 4$ as termination parameter (when alle boxes of the list have a width less than ϵ, the computation stops).

After 175 iterations of Alg. 2 the computations terminated with a lower bound of f^* of $-3.7252E-10$ and an upper bound of $1.8626E-10$. The global minimizer was included in 4 boxes whose union was $([-.6104E-04, .6104E-04], [-.6104E-04, .6104E-04])^T$.

The algorithm required $N = 686$ inclusion function evaluations where 415 evaluations were used to construct inclusions for the objective function and 271 were used to construct inclusions for the constraint function.

By comparison, the tunnelling algorithm in Levy-Gomez (1985) required $N_f = 12495$ and $N_c = 16704$ where N_f is the number of objective function evaluations and N_c is the number of constraint function evaluations which were neeeded for 20 runs of this problem. The 20 starting points were randomly chosen. Let $L_x(x, u, v)$ be the derivative of the Lagrangian function of the problem with respect to x, that is the vector which occurs as the left-hand side of the first line of (1.4). Then a point x is accepted as a local minimizer in Levy-Gomez when

$$L_x(x, u, v)^T L_x(x, u, v) \leq 1.0E - 9.$$

(Originally, Levy-Gomez used an augmented Lagrangian function instead of L. For simplicity, we provide the above acceptance criterion which does not deviate essentially from the one of Levy-Gomez.)

Example 2. (Levy-Gomez (1985), p. 243, Problem 2). This example is Example 1 solved using Alg. 2 with the preprocessing provided by Alg. 3.

The input data were the same as for Example 1.

After 1 iteration of Alg. 3 a feasible point, the zero vector, was found. With this initial point Alg. 2 required 114 iterations before it terminated with a lower bound of f^* of 0.0 and an upper bound of

$1.8626E-10$. The global minimizer was included in 4 boxes whose union was $([-.6104E-04, .6104E-04], [-.6104E-04, .6104E-04])^T$.

The algorithm required $N = 351$ inclusion function evaluations where 256 evaluations were used to construct inclusions for the objective function and 95 were used to construct inclusions for the constraint function.

Comparison data were given in Example 1.

Example 3. (Levy-Gomez(1985), p. 243, Problem 2). Again we consider the constrained problem of Exs. 1 and 2. Here, Alg. 2 combined with the interval Newton method and the monotonicity test was applied as described on page 195. That is, if the box treated was feasible, each iteration of Alg. 2 was supplemented by the monotonicity test and by one iteration of the interval Newton algorithm, as it was the case with the accelerated treatment of unconstrained problems, cf. page 125 ff.

The input data was the same as for Exs. 1 and 2, with the exception that, additionally, the algorithm was supplied with one feasible point, which was already known from Ex. 2.

Alg. 2 required 62 iterations before it terminated delivering about the same output data as in Ex. 2. The algorithm needed 166 inclusion function evaluations where 134 evaluations were used to compute inclusions for the objective function and 32 were used to compute inclusions for the constrained function.

Example 4. (Zowe (1985), p. 350). We consider the following two-dimensional problem:

$$\text{minimize } f(x) := (x_1 - 2)^2 + (x_2 - 1)^2$$

s. t.

$$g_1(x) := x_1^2 - x_2 \leq 0,$$
$$g_2(x) := x_1 + x_2 - 2 \leq 0$$

for $x \in \mathbf{R}^2$. The global minimizer is $x^* = (1,1)^T$ which is an edge point of the feasible domain. Both constraints are active. The global minimum is $f^* = 1$.

We transformed the problem to an unconstrained but nonsmooth problem by using the penalty function (1.6). This penalty approach was combined with Alg. 1 in order to discard infeasible areas. Such a combination makes it unnecessary to use a sequence of penalty factors ρ_n to get convergence, cf. Sec. 1.4, because Alg. 1 causes the necessary push of the solution into the feasible domain. The algorithm for unbounded domains (Alg. 1 of Sec. 4.2) was then applied, incorporating the monotonicity test, to enclose the solutions of the problem. As inclusion function we took natural interval extensions and, for smaller boxes, meanvalue forms.

The input data were:

$X = \overline{\mathbf{R}}^2$ as starting domain,
$\lambda = 5$ as boundary parameters, cf. Sec. 4.2,
$\epsilon = E - 6$ as termination parameter (when alle boxes of the list have a width less than ϵ, the computation stops),
ρ $= 1, 1.1, 2, 10, 100, 1000, 10000$ as penalty factors.

The output data are described using

N the number of inclusion function evaluations of (1.6),
N_b the number of final boxes (they are all of width smaller than ϵ and their union includes $X^* = \{x^*\}$).

The following results were obtained as a function of the penalty factor ρ, where f^* was included in an interval of width $5.97E - 7$ in each case:

ρ	N	N_b
1	252	3
1.1	260	3
2	340	3
10	384	6
100	390	6
1000	390	6
10000	390	6

Bibliography

Aird, T.J., Rice, J.R. (1977). Systematic search in higher dimensional sets, *SIAM Journal on Numerical Analysis*, **14**, pp. 296-312.

Albrecht, R., Kulisch, U. (eds.) (1977). *Grundlagen der Computer-Arithmetik*, Computing Supplementum 1, Springer-Verlag, Vienna.

Albrycht, J., Wisniewski, H. (eds.) (1985). *Proc. Polish Symp. Interval and Fuzzy Math.*, Inst. Math., Tech. Univ. Poznan.

Alefeld, G. (1968). Intervallrechnung über den komplexen Zahlen und einige Anwendungen, Thesis, Universität Karlsruhe.

Alefeld, G. (1984). On the convergence of some interval-arithmetic modifications of Newton's method, *SIAM Journal on Numerical Analysis*, **21**, pp. 363-372.

Alefeld, G., Grigorieff, R.D. (eds.) (1979). *Fundamentals of Numerical Computation (Computer-Oriented Numerical Analysis)*, Computing Supplementum 2, Springer-Verlag, Vienna.

Alefeld, G., Herzberger, J. (1974). *Einführung in die Intervallrechnung*, Bibliographisches Institut, Mannheim.

Alefeld, G., Herzberger, J. (1983). *Introduction to Interval Computations*, Academic Press, New York.

Alefeld, G., Lohner, R. (1985). On higher order centered forms, *Computing*, **35**, pp. 177-184.

Alefeld, G., Platzöder, L. (1983). A quadratically convergent Krawczyk-like algorithm, *SIAM Journal on Numerical Analysis*, **20**, pp. 210-219.

Alefeld, G., Rokne, J. (1981). On the interval evaluation of rational functions in interval arithmetic, *SIAM Journal on Numerical Analy-*

sis, **18**, pp. 862-870.

Anderssen, R.S., Bloomfield, P. (1975). Properties of the random search in global optimization, *J. Optim. Theor. and Appl.*, **16**, pp. 383-398.

Apostolatos, N., Kulisch, U., Krawczyk, R., Lortz, B., Nickel, K., Wippermann, H.W. (1968). The algorithmic language TRIPLEX ALGOL 60, *Numerische Mathematik*, **11**, pp. 75-180.

Archetti, F., Betro, B. (1979). A probabilistic algorithm for global optimization, *Calcolo*, **16**, pp. 335-343.

Armstrong, R., Charnes, A., Phillips, F. (1979). Page cuts for integer interval linear programming, *Discrete Applied Math.*, **1**, pp. 1-14.

Asaithambi, N.S., Shen, Z., Moore, R.E. (1982). On computing the range of values, *Computing*, **28**, pp. 225-237.

Baba, N. (1979). Global optimization of functions by the random optimization method, *Int. J. Control (O. B.)*, **30**, pp. 1061-1066.

Bachem, A., Grötschel, M., Korte, B. (eds.) (1983). *Mathematical Programming: The State of the Art*, Springer-Verlag, Berlin.

Bao, P.G., Rokne, J.G. (1987). Existence of a unique zero of nonlinear systems, Preprint.

Basso, P. (1982). Iterative methods for the localisation of the global maximum, *SIAM Journal on Numerical Analysis*, **19**, pp. 781-792.

Basso, P. (1985). Optimal search for the global maximum of functions with bounded seminorm, *SIAM Journal on Numerical Analysis*, **22**, pp. 888-903.

Baumann, E. (1986). Globale Optimierung stetig differenzierbarer Funktionen einer Variablen, *Freiburger Intervall-Berichte* 86/6, Institut für Angewandte Mathematik, Universität Freiburg.

Baumann, E. (1987). Optimal centered forms, *Freiburger Intervall-Berichte* 87/3, Institut für Angewandte Mathematik, Universität Freiburg, pp. 5-21.

Bazaraa, M.S., Jarvis, J.J. (1977). *Linear Programming and Network Flows*, Wiley, New York.

Bazaraa, M.S., Shetty, C.M. (1979). *Nonlinear Programming Theory and Algorithms*, Wiley, New York.

Beckeu, R.W., Lago, G.V. (1970). A global optimization algorithm, *Proceedings of the 8th Annual Allerton Conference on Circuit and System Theory*, IEEE, New York, pp. 3-12.

Benson, H.P. (1982). On the convergence of two branch-and-bound algorithms for nonconvex programming problems, *J. Optim. Theor. and Appl.*, **36**, pp. 129-134.

Bertsekas, D.P. (1982). *Constrained Optimization and Lagrange Multiplier Methods*, Academic Press, New York.

Blum, E., Oettli, W. (1975). *Mathematische Optimierung*, Springer-Verlag, Berlin.

Boggs, P.T., Byrd, R.H., Schnabel, R.B. (eds.) (1985). *Numerical Optimization 1984*, SIAM, Philadelphia.

Bohlender, G., Böhm, H., Kaucher, E., Kirchner, R., Kulisch, U., Rump, S., Ullrich, Ch., von Gudenberg, W. (1981). PASCAL-SC: A Pascal for Contemporary Scientific Computation, IBM Report RC 9009.

Brayton, R.K., Cullum, J. (1979). An algorithm for minimizing a differentiable function subject to box constraints and errors, *J. Optim. Theor. and Appl.*, **29**, pp. 521-58.

Brent, R.P. (1973). *Algorithms for Minimization Without Derivatives*, Prentice-Hall, Englewood Cliffs, New Jersey.

Caprani, O., Madsen, K. (1979). Interval methods for global optimization, Report No. NI 79-09, Institute for Numerical Analysis, Technical University of Denmark.

Caprani, O., Madsen, K. (1980). Mean value forms in interval analysis, *Computing*, **25**, pp. 147-154.

Cargo, G.T., Shisha, O. (1966). The Bernstein form of a polynomial,

Journal of Research of the National Bureau of Standards, Section B, **70B**, pp. 79-81.

Chabrillac, Y., Crouzeix, J.-P. (1987). Continuity and differentiability properties of monotone real functions of several real variables, *Mathematical Programming Study*, **30**, pp. 1-16.

Clarke, F.H. (1983). *Optimization and Nonsmooth Analysis*, Wiley, New York.

Cohn, D.A., Potter, J.B., Ginsberg, M. (1979). Implementation and evaluation of interval arithmetic software, Report 5, The CDC CYBER 70 Systems Technical Report 0-79-1, U. S. Army Engineer, Waterways Experiment Station.

Conte, S.D., de Boor, C. (1980). *Elementary Numerical Analysis*, 3rd ed. McGraw-Hill, New York.

Cornelius, H., Lohner, R. (1984). Computing the range of values of real functions with accuracy higher than second order, *Computing*, **33**, pp. 331-347.

Crippen, G.M. (1975). Global optimization and polypeptide conformation, *J. Comput. Phys.*, **18**, pp. 224-231.

Daniel, J.W. (1973). Newton's method for nonlinear inequalities, *Numerische Mathematik*, **21**, pp. 381-387.

Dantzig, G., Eaves, B. (1973). Fourier-Motzkin elimination and its dual, *J. Combinatorial Theory (A)*, **14**, pp. 288-297.

Dantzig, G., Eaves, B. (1975). Fourier-Motzkin elimination and its dual with application to integer programming. In: Roy, D. (ed.), *Combinatorial programming: methods and applications*, Proceedings of the NATO Advanced Study Institute, D. Reidel Publishing Company, Dordrecht, Holland, pp. 93-102.

Davidon, W.C. (1959). Variable metric methods for minimization, *A.E.C. Res. and Develop. Rep.* **ANL-5990**, Argonne National Laboratory, Argonne, Illinois.

Davidon, W.C. (1980). Conic approximations and collinear scalings for optimizers, *SIAM Journal on Numerical Analysis*, **17**, pp. 268-281.

Demyanov, V.F. (1986). Quasidifferentiable functions: necessary conditions and descent directions, *Mathematical Programming Study*, **29**, pp. 20-43.

Demyanov, V.F., Dixon, L.C.W. (eds.) (1986). Quasidifferential calculus, *Mathematical Programming Study*, **29**, North-Holland, Amsterdam.

Demyanov, V.F., Pallaschke, D. (eds.) (1985). *Nondifferentiable Optimization: Motivations and Applications*, Proceedings, Sopron, Hungary, 1984. Springer-Verlag.

Demyanov, V.F., Polyakova, L.N., Rubinov, A.M. (1986). Nonsmoothness and quasidifferentiability, *Mathematical Programming Study*, **29**, pp. 1-19.

Demyanov, V.F., Vasilev, L.V. (1985). *Nondifferentiable Optimization*, Optimization Software, Inc., New York.

Dennis, J.E. (1978). A brief introduction to quasi-Newton methods. In: Golub, G.H., Oliger, J. (eds.), *Numerical Analysis*, Amer. Math. Soc., Providence, RI, pp. 19-52.

Dennis, J.E., Moré, J.J. (1977). Quasi-Newton methods, motivation and theory, *SIAM Review*, **21**, pp. 443-459.

Dennis, J.E., Schnabel, R.B. (1983). *Numerical Methods for Unconstrained Optimization and Nonlinear Equations*, Prentice-Hall, Englewood Cliffs, NJ.

Devroye, L.P. (1978). Progressive global random search of continuous functions, *Mathematical Programming*, **15**, pp. 330-342.

Dinkel, J.J., Tretter, M.J., Wong, D. (1987). Interval Newton methods and perturbed problems. In: Moore (ed.) (1988).

Dixon, L.C.W. (1978). Global optima without convexity. In: Greenberg, H. (ed.) *Design and Implementation Optimization Software*, Sijthoff and Noordhoff, Alphen aan den Rijn, pp. 449-479.

Dixon, L.C.W., Szegö, G.P. (eds.) (1975). *Towards Global Optimiza-*

tion, North-Holland, Amsterdam.

Dixon, L.C.W., Szegö, G.P. (eds.) (1977). *Towards Global Optimization*, **2**, North-Holland, Amsterdam.

Duff, I.S. (1987). The influence of vector and parallel processors on numerical analysis. In: Iserles-Powell (eds.) (1987), pp. 359-408.

Duffin, R.J. (1974). On Fourier's analysis of linear inequality systems, *Mathematical Programming Study*, **1**, pp. 71-95.

Dussel, R. (1972). Einschliessung des Minimalpunktes einer streng konvexen Funktion auf einem n-dimensionalen Quader, Dissertation, Universität Karlsruhe, Karlsruhe.

Dussel, R. (1975). Einschliessung des Minimalpunktes einer streng konvexen Funktion auf einem n-dimensionalen Quader. In: Nickel (ed.) (1975), pp. 169-177.

Dussel, R., Schmitt, B. (1970). Die Berechnung von Schranken für den Wertebereich eines polynoms in einem Intervall, *Computing*, **6**, pp. 35-60.

Evtushenko, Y. (1971). Numerical methods for finding global extrema (case of a uniform mesh), *USSR Comp. Math. and Math. Phys.*, **11**, pp. 38-64.

Evtushenko, Y. (1985). *Numerical Optimization Techniques*, Optimization Software Inc., New York.

Fan, Ky (1956). On systems of linear inequalities. In: Kuhn, H.W., Tucker, A.W. (eds.), *Linear Inequalities and Related Systems. Annals of Math. Studies*, **39**, pp. 96-156.

Fiacco, A.V., McCormick, G.P. (1968). *Nonlinear Programming: Sequential Unconstrained Minimization Techniques*, Wiley, New York.

Findler, N.V., Lo, C., Lo, L. (1987). Pattern search for optimization, *Mathematics and Computers in Simulation*, **29**, pp. 41-50.

Fletcher, R. (1980). *Practical Methods of Optimization, Vol. 1, Unconstrained Optimization*, Wiley, New York.

Fletcher, R. (1981). *Practical Methods of Optimization, Vol. 2, Constrained Optimization*, Wiley, New York.

Fletcher, R. (1983). Penalty functions. In: Bachem-Grötschel-Korte (eds.) (1983), pp. 87-114.

Fletcher, R. (1985). An l_1 penalty method for nonlinear constraints. In: Boggs-Byrd-Schnabel (eds.) (1985), pp. 26-40.

Fletcher, R. (1987a). Recent developments in linear and quadratic programming. In: Iserles-Powell (eds.) (1987), pp. 213-244.

Fletcher, R. (1987b). *Practical Methods of Optimization*, Wiley, New York.

Fletcher, R., Reeves, C.M. (1964). Function minimization by conjugate gradients, *Comp. J.*, **7**, pp. 149-154.

Fujii, Y., Ichida, K., Kiyono, T. (1977). A method for finding the greatest value of a multivariable function using interval arithmetic, *Inf. Process. Soc. Jpn. (Joho Shori) (Japan)*, **18**, pp. 1095-1110.

Fuijii, Y., Ichida, K., Ozasa, M. (1986). Maximization of multivariate functions using interval analysis. In: Nickel (ed.) (1986), pp. 37-56.

Garcia-Palomares, U.M., Restuccia, A. (1981). A global quadratic algorithm for solving a system of mixed equalities and inequalities, *Mathematical Programming*, **21**, pp. 290-300.

Gill, P.E., Murray, W. (1974). *Numerical Methods for Constrained Optimization*, Academic Press, New York.

Gill, P.E., Murray, W. (1978). Algorithms for the solution of the nonlinear least-squares problems, *SIAM Journal on Numerical Analysis*, **15**, pp. 977-992.

Gill, P.E., Murray, W. (1980). A numerical investigation of ellipsoid algorithms for large-scale linear programming, *Systems Optimization Laboratory*, Report SOL 80-27, Stanford University, Stanford, California.

Gill, P.E., Murray, W., Wright, M.H. (1981). *Practical Optimization*, Academic Press, New York.

Goffin, J.L. (1977). On convergence rate of subgradient optimization, *Math. Progr.*, **13**, pp. 329-347.

Golden, B.L., Alt, F.B. (1979). Interval estimation of a global optimum for large combinatorial problems, *Naval Research Log. Quarterly*, **26**, pp. 69-77.

Goldfeldt, S.M., Quandt, R.E., Trotter, H.F. (1966). Maximization by quadratic hill-climbing, *Econometrica*, **34**, pp. 541-551.

Hansen, E.R. (1965). Interval arithmetic in matrix computations, part I, *SIAM Journal on Numerical Analysis*, **2**, pp. 308-320.

Hansen, E.R. (1968). On solving systems of equations using interval arithmetic, *Math. Comput.*, **22**, pp. 374-384.

Hansen, E.R. (ed.) (1969a). *Topics in Interval Analysis*, Oxford University Press, Oxford.

Hansen, E.R. (1969b). The centered form. In: Hansen (ed.) (1969a), pp. 102-105.

Hansen, E.R. (1969c). On the solution of linear algebraic equations with interval coefficients, *Linear Algebra and Appl.*, **2**, pp. 153-165.

Hansen, E.R. (1978a). Interval forms of Newton's method, *Computing*, **20**, pp. 153-163.

Hansen, E.R. (1978b). A globally convergent interval analytic method for computing and bounding real roots, *BIT*, **18**, pp. 415-424.

Hansen, E.R. (1979). Global optimization using interval analysis - the one-dimensional case, *J. Optim. Theor. and Appl.*, **29**, pp. 331-344.

Hansen, E.R. (1980). Global optimization using interval analysis - the multidimensional case, *Numerische Mathematik*, **34**, pp. 247-270.

Hansen, E.R. (1984). Global optimization with data perturbations, *Computers and Oper. Res.*, **11**, pp. 97-104.

Hansen, E.R. (1988). An overview of global optimization using interval analysis. In: Moore (ed.) (1988).

Hansen, E.R., Greenberg, R.I. (1983). An interval Newton method,

Applied Math. and Comp., **12**, pp. 89-98.

Hansen, E.R., Sengupta, S. (1980). Global constrained optimization using interval analysis. In: Nickel (ed.) (1980), pp. 25-47.

Hansen, E.R., Sengupta, S. (1981). Bounding solutions of systems of equations using interval analysis, *BIT*, **21**, pp. 203-211.

Hansen, E.R., Sengupta, S. (1983). Summary and steps of a global nonlinear constrained optimization algorithm, LMSC-D889778, Lockheed Missiles and Space Co., Sunnyvale, California.

Hansen, E.R., Smith, R.R. (1967). Interval arithmetic in matrix computations, part II, *SIAM Journal on Numerical Analysis*, **4**, pp. 1-9.

Hansen, E.R., Walster, G.W. (1982). Global optimization in nonlinear mixed integer problems, *Proceedings IMACS World Congress Montreal*, Vol. 1, pp. 379-381.

Hansen, E.R., Walster, G.W. (1987a). Nonlinear equations and optimization, Preprint.

Hansen, E.R., Walster, G.W. (1987b). Bounds for Lagrange multipliers and optimal points, Preprint.

Hanson, R.J. (1968). Interval arithmetic as a closed arithmetic system on a computer, Jet Propulsion Laboratory Report 197, Pasadena, California.

Hart, J.F., *et al.* (1968). *Computer Approximations*, Wiley, New York.

Hartman, J.K. (1973a). Some experiments in global optimization, *Naval Research Log. Quarterly*, **20**, pp. 569-576.

Hartman, J.K. (1973b). A new method for global optimization, Technical Report NPS55HH73041A, Naval Post-Graduate School.

Hestenes, M.R. (1975). *Optimization Theory: The Finite Dimensional Case*, Wiley, New York.

Hiebert, K.L. (1980). Solving systems of linear equations and inequalities, *SIAM Journal on Numerical Analysis*, **17**, pp. 447-464.

Himmelblau, D. (1972). *Applied Nonlinear Programming*, McGraw-Hill, Inc., New York.

Hiriart, J.-B. (1986). When is a point x satisfying $\nabla f(x) = 0$ a global minimum of f, *Amer. Math. Monthly*, **93**, pp. 556-558.

Horst, R. (1979). *Nichtlineare Optimierung*, Hauser-Verlag, München.

Horst, R. (1985) Globally convergent methods in multiextremal global optimization, Preprint.

Horst, R. (1986). A general class of branch and bound methods in global optimization with some new approaches for concave minimization, *J. Optim. Theor. and Appl.*, **51**, pp. 271-291.

Horst, R., Tuy, H. (1987). On the convergence of global methods in multiextremal optimization, *J. Optim. Theor. and Appl.*, **54**, pp. 253-271.

Ichida, K., Fujii, Y. (1979). An interval arithmetic method for global optimization, *Computing*, **23**, pp. 85-97.

Iserles, A., Powell, M.J.D. (eds.) (1987). *The State of the Art in Numerical Analysis*, Proceedings of the Joint IMA (SIAM) Conference at University of Birmingham, 1986, Clarendon Press, Oxford.

Jacobsen, S.E., Torabi, M. (1978). A global minimization algorithm for a class of one-dimensional functions, *J. Math. Anal. and Appl.*, **62**, pp. 310-324.

Jacoby, S.L.S., Kowalik, J.S., Pizzo, J.T. (1972). *Iterative Methods for Nonlinear Optimization Problems*, Prentice-Hall, Englewood Cliffs, NJ.

Kagiwada, H., Kalaba, R., Rasakhoo, N., Spingarn, K. (1986). *Numerical Derivatives and Nonlinear Analysis*, Plenum Press, New York.

Kahan, W.M. (1968). A more complete interval arithmetic, Lecture notes for a summer course at the University of Michigan.

Kamel, K., Araki, Y., Inoue, K. (1979). Heuristic attainment in multimodal maximum searching problems, *Mem. Res. Inst. Sc. and Eng. Ritsumeikan Univ.*, **36**, pp. 23-30.

Kearfott, R.B. (1987). Abstract generalized bisection and a cost bound, *Mathematics of Computation*, **49**, pp. 187-202.

Kiefer, J. (1957). Optimum search and approximation methods under minimum regularity assumptions, *SIAM Journal*, **5**, pp. 105-136.

Kolev, L.V. (1984). Global constrained optimization via interval analysis technique, Proceedings of the International AMSC Conf. Modelling and Simulation, Athens, Vol. 1.2, pp. 175-188.

Krawczyk, R. (1969). Newton-Algorithmen zur Bestimmung von Nullstellen mit Fehlerschranken, *Computing*, **4**, pp. 187-201.

Krawczyk, R. (1975). Fehlerabschatzung bei linearen Optimierung. In: Nickel (ed.) (1975), pp. 215-228.

Krawczyk, R. (1982). Zentrische Formen und Intervalloperatoren, *Freiburger Intervall-Berichte* 82/1, Institut für Angewandte Mathematik, Universität Freiburg.

Krawczyk, R. (1983). Intervallsteigungen für rationale Funktionen und zugeordnete zentrische Formen, *Freiburger Intervall-Berichte* 83/2, Institut für Angewandte Mathematik, Universität Freiburg.

Krawczyk, R. (1986). A class of interval Newton operators, *Computing*, **37**, pp. 179-183.

Krawczyk, R., Neumaier, A. (1985). Interval slopes for rational functions and associated centered forms, *SIAM Journal on Numerical Analysis*, **22**, pp. 604-616.

Krawczyk, R., Nickel, K. (1982). Die zentrische Form in der Intervallarithmetik, ihre quadratische Konvergenz und ihre Inklusionsisotonie, *Computing*, **28**, pp. 117-132.

Kulisch, U., Miranker, W.L. (1981). *Computer Arithmetic in Theory and Practice*, Academic Press, New York.

Kulisch, U., Ullrich, C., (eds.) (1982). *Wissenschaftliches Rechnen und Programmiersprachen*, Teubner, Stuttgart.

Laveuve, S.E. (1975). Definition einer Kahan-Arithmetik und ihre Implementierung. In: Nickel (ed.) (1975), pp. 236-245.

Lebourg, G. (1975). Valeur moyenne pour gradients, *Co. R. Acad. Sci.*, Paris 281, pp. 795-797.

Lemaréchal, C. (1980). Nondifferentiable optimization. In: Dixon, Spedicato, Szegö (eds.), *Nonlinear Optimization, Theory and Algorithms*, Birkhäuser, Boston.

Levenberg, K. (1944). A method for the solution of certain problems in least squares, *Quart. Appl. Math.*, **2**, pp. 164-168.

Levy, A.V., Gomez, S. (1985). The tunneling method applied to global optimization. In: Boggs-Byrd-Schnabel (eds.) (1985), pp. 213-244.

Levy, A.V., Montalvo, A. (1985). The tunnelling algorithm for the global minimization of functions, *SIAM J. Sci. Stat. Comput.*, **6**, pp. 15-29.

Mancini, L.J. (1975). Applications of interval arithmetic in signomial programming, Technical Report SOL 75-23, Stanford University, Stanford, California.

Mancini, L.J., McCormick, G.P. (1976). Bounding global minima, *Math. Oper. Res.*, **1**, pp. 50-53.

Mancini, L.J., McCormick, G.P. (1979). Bounding global minima with interval arithmetic, *Math. Oper. Res.*, **27**, pp. 743-754.

Mangasarian, O.L. (1969). *Nonlinear Programming*, McGraw-Hill, New York.

Marquardt, D. (1963). An algorithm for least-squares estimation of non-linear parameters, *SIAM J. Appl. Math.*, **11**, pp. 431-441.

McCormick, G.P. (1972). Attempts to calculate global solutions of problems that may have local minima. In: Lootsma, F.A. (ed.), *Numerical Methods for Nonlinear Optimization*, Academic Press, London.

McCormick, G.P. (1976). Computability of global solutions to factorable nonconvex programs: Part I - Convex underestimating problems, *Mathematical Programming*, **10**, pp. 147-175.

McCormick, G.P. (1980). Locating an isolated global minimizer of a constrained non-convex program, *Math. Oper. Res.*, **5**, pp. 435-443.

McCormick, G.P. (1981). Finding the global minimum of a function of one variable using the methods of constant signed higher order derivatives. In: Mangasarian, O.L, Meyer, R.R., Robinson, S.M. (eds.), *Nonlinear Programming 4, Proceedings Symposium* Madison, July 14-16, Academic Press, New York, pp. 223-243.

McCormick, G.P. (1983). *Nonlinear Programming: Theory, Algorithms and Applications*, Wiley, New York.

McCormick, G. P. (1985). Global solution to factorable nonlinear optimization using separable programming techniques, NBSIR 85-8206, U.S. Dept. of Commerce.

Mentzer, S.G. (1986). Potent interval bounds for deterministic global optimization, Preprint.

Mifflin, R. (1977). An algorithm for constrained optimization with semismooth functions, *Math. Oper. Res.*, **2**, pp. 191-207.

Miller, W. (1972a). On an interval-arithmetic matrix method, *BIT*, **12**, pp. 213-219.

Miller, W. (1972b). Quadratic convergence in interval arithmetic, Part II, *BIT*, **12**, pp. 291-298.

Miller, W., Chuba, W. (1972). Quadratic convergence in interval arithmetic, Part I, *BIT*, **12**, pp. 284-290.

Minoux, M. (1986). *Mathematical Programming, Theory and Algorithms*, Wiley, New York.

Mladineo, R.H. (1986). An algorithm for finding the global maximum of a multimodal, multivariate function, *Mathematical Programming*, **34**, pp. 188-200.

Mockus, J. (1975). On the Bayes methods for seeking the extremal point. In: *Proceedings of the 6th Triennial World Congress of the International Federation of Automatic Control, Pittsburgh: ISA*, 30.3/1-4.

Mockus, J. (1976). On the Bayes methods for seeking the extremal point, *International Symposium of Information Theory*, IEEE, New York.

Mohd, I.B. (1986). Global optimization using interval arithmetic, Ph.D. Thesis, University of St Andrews.

Moore, R.E. (1959). Automatic error analysis in digital computation, Technical Report LMSD-4842, Lockheed Missiles and Space Division, Sunnyvale, California.

Moore, R.E. (1962). Interval arithmetic and automatic error analysis in digital computation, Ph.D. Thesis, Stanford University.

Moore, R.E. (1966). *Interval Analysis*, Prentice-Hall, Englewood Cliffs, NJ.

Moore, R.E. (1969). *Intervallanalyse*, Oldenburg-Verlag, München.

Moore, R.E. (1976). On computing the range of values of a rational function of n variables over a bounded region, *Computing*, **16**, pp. 1-15.

Moore, R.E. (1977). A test for existence of solutions to non-linear systems, *SIAM Journal on Numerical Analysis*, **14**, pp. 611-615.

Moore, R.E. (1978). A computational test for convergence of iterative methods for non-linear systems, *SIAM Journal on Numerical Analysis*, **15**, pp. 1194-1196.

Moore, R.E. (1979). *Methods and Applications of Interval Analysis*, SIAM, Philadelphia.

Moore, R.E. (1980). Interval methods for nonlinear systems, *Computing Suppl.*, **2**, pp. 113-120.

Moore, R.E. (1988) (ed.). *Reliability in Computing: The Role of Interval Methods*, Academic Press, New York. Forthcomming.

Moore, R.E., Jones, S.T. (1977). Safe starting regions for iterative methods, *SIAM Journal on Numerical Analysis*, **14**, pp. 1051-1065.

Moore, R.E., Ratschek, H. (1987). Inclusion functions and global optimization II, *Mathematical Programming*. Forthcoming.

Moré J.J., Cosnard, M.Y. (1979). Numerical solution of nonlinear equations, *ACM Transactions on Mathematical Software*, **5**, pp. 64-85.

Moré, J.J., Garbow, B.S., Hillstrom, E. (1981). Testing unconstrained optimization software, *ACM Transactions on Mathematical Software*, **7**, pp. 17-41.

Moré, J.J., Sorensen, D.C. (1979). On the use of directions of negative curvature in a modified Newton method, *Mathematical Programming*, **16**, pp. 1-20.

Morel, E., Renvoise, C. (1979). Global optimization by suppression of partial redundancies, *Comm. ACM*, **22**, pp. 96-103.

Neumaier, A. (1984). Interval iteration for zeroes of systems for equations, *Freiburger Intervall-Berichte* 84/4, Institut für Angewandte Mathematik, Universität Freiburg, pp. 29-58.

Neumaier, A. (1985). Interval iteration for zeroes of systems of equations, *BIT*, **25**, pp. 256-273.

Neumaier, A. (1988). The enclosure of parameter-dependent systems of equations. In: Moore (ed.) (1988).

Nickel, K. (1971). On the Newton method in interval analysis, *Mathematics Research Center Report 1136*, University of Wisconsin.

Nickel, K. (ed.) (1975). *Interval Mathematics*, Proceedings of the International Symposium, Karlsruhe 1975, Springer-Verlag, Vienna.

Nickel, K. (1977). Die Überschätzung des Wertebereichs einer Funktion in der Intervallrechnung mit Anwendungen auf lineare Gleichungssysteme, *Computing*, **18**, pp. 15-36.

Nickel, K. (ed.) (1980). *Interval Mathematics 1980*, Proceedings of the International Symposium, Freiburg 1980, Academic Press, New York.

Nickel, K. (ed.) (1986). *Interval Mathematics 1985*, Proceedings of the International Symposium, Freiburg 1985, Springer Verlag, Wienna.

Nickel, K., Ritter, K. (1972). Termination criterion and numerical convergence, *SIAM Journal on Numerical Analysis*, **9**, pp. 277-283.

Oelschlägel, D., Süsse, H. (1980). Interval analytic treatment of convex programming problems, *Computing*, **24**, pp. 213-225.

Oelschlägel, D., Süsse, H. (1981). Optimierung mit Hilfe des Intervall-Newton-Verfahrens, *ZAMM*, **61**, pp. 243-251.

Oelschlägel, D., Süsse, H. (1983). Berechnung von Intervall-Hüllen in der Optimierung, *ZAMM*, **63**, pp. 379-386.

Oelschlägel, D., Süsse, H., Schmielow, R. (1979). Die Behandlung von Aufgaben aus der Optimierung dynamischer Systeme mittels intervallanalytischer Methoden, *Wiss. Z. Leuna-Merseburg*, **21**, pp. 105-111.

Ortega, J.M., Rheinboldt, W.C. (1970). *Iterative Solution of Nonlinear Equations in Several Variables*, Academic Press, New York.

Pardalos, P.M. (1986). Constrained global optimization references, CS-86-02, Dept. of Comp. Sci., The Pennsylvania State University, University Park, Pennsylvania.

Pardalos, P.M., Rosen, J.B. (1987). *Constrained Global Optimization: Algorithms and Applications*, Springer-Verlag, Berlin.

Parviainen, S. (1985). A direct random search method for global optimization, *IIASA workshop on global optimization*, Sopron.

Pfranger, R. (1970). A heuristic method for global optimization, *Unternehmensforschung*, **14**, pp. 27-50.

Pierre, D.A., and Lowe, M.J. (1975). *Mathematical Programming via Augmented Lagrangians*, Addison-Wesley, New York.

Piyavskii, S.A. (1972). An algorithm for finding the absolute extremum of a function, *USSR Comput. Math. Phys.*, **12**, pp. 57-67.

Polak, E., Ribiére, G. (1969). Note sur la convergence de methodes de directions conjugées, *Rev. Franc. Informat. Recherche Opérationelle*, **16**, pp. 35-43.

Powell, M.J.D. (1964) An efficient method for finding the minimum of a function of several variables without calculating derivatives, *The Computer Journal*, **7**, pp. 155-162.

Powell, M.J.D. (1982). *Nonlinear Optimization 1981*, Academic Press, London.

Powell, M.J.D. (1983). Variable metric methods for constrained optimization. In: Bachem-Grötschel-Korte (eds.) (1983), pp. 288-311.

Powell, M.J.D. (1986). Convergence properties of algorithms for nonlinear optimization, *SIAM Review*, **28**, pp. 487-500.

Powell, M.J.D. (1987a). Updating conjugate directions by the BFGS formula, *Math. Prog.*, **38**, pp. 29-46.

Powell, M.J.D. (1987b). Methods for nonlinear constraints in optimization, In: Iserles-Powell (eds.) (1987), pp. 325-358.

Price, W.L. (1977). A controlled random search procedure for global optimisation, *Computer J.*, **20**, pp. 1375-1383.

Qi, L. (1982). A note on the Moore test for nonlinear systems, *SIAM Journal on Numerical Analysis*, **19**, pp. 851-857.

Quan, Z., Bai-Chuan, J., Song-Lin, A. (1978). A method for searching global minimum, *ACTA Math. Appl. Sin. (China)*, **1**, pp. 161-174.

Rall, L.B. (1981). *Automatic Differentiation: Techniques and Applications*, Notes in Computer Science No. 120, Springer, New York.

Rall, L.B. (1983). Mean value and Taylor forms in interval analysis, *SIAM Journal on Mathematical Analysis*, **14**, pp. 223-238.

Rall, L.B. (1985). Global optimization using automatic differentiation and interval iteration, Technical Summary Report No. 2832, Mathematics Research Center, University of Wisconsin, Madison.

Rall, L.B. (1986). Improved interval bounds for the range of functions. In: Nickel (ed.) (1986), pp. 143-155.

Ratschek, H. (1975). Nichtnumerische Aspekte der Intervallarithmetik. In: Nickel (ed.) (1975), pp. 48-74.

Ratschek, H. (1976). Ueber den Quasilinearen Raum, *Berichte Math.-*

Statist. Sekt., Forschungszentrum Graz No. 65.

Ratschek, H. (1978). Zentrische Formen, *Zeitschrift für Angewandte Mathematik und Mechanik*, **58**, pp. T434-T436.

Ratschek, H. (1980). Centered forms, *SIAM Journal on Numerical Analysis*, **17**, pp. 656-662.

Ratschek, H. (1982). Intervallarithmetik als Algebraische Theorie, *Berichte Math.-Statist. Sekt.*, Forschungszentrum Graz No. 189.

Ratschek, H. (1985a). Inclusion functions and global optimization, *Mathematical Programming*, **33**, pp. 300-317.

Ratschek, H. (1985b). Interval tools for global optimization, *IIASA workshop on global optimization*, Sopron.

Ratschek, H. (1988). Interval mathematics. In: Holzman, A.G., Kent, A., Williams, J.G. (eds.), *Encyclopedia of Computer Science and Technology*, Marcel Dekker, New York. Forthcoming.

Ratschek, H., Rokne, J. (1984). *Computer Methods for the Range of Functions*, Ellis Horwood, Chichester, 1984.

Ratschek, H., Rokne, J. (1987). Efficiency of a global optimization algorithm, *SIAM Journal on Numerical Analysis*, **24**, pp. 1191-1201.

Ratschek, H., Sauer, N. (1982). Linear interval equations, *Computing*, **28**, pp. 105-115.

Ratschek, H., Schröder, G. (1981). Centered forms for functions in several variables, *Journal of Mathematical Analysis and Applications*, **82**, pp. 543-552.

Ratschek, H., Voller, R.L. (1988). Unconstrained optimization over unbounded domains, Preprint.

Rinnooy Kan, A.H.G., Timmer, G.T. (1986). Global optimization, Rep. 8612/A, Econometric Institute, Erasmus University, Rotterdam.

Rivlin, T.J. (1970). Bounds on a polynomial, *Journal of Research of the National Bureau of Standards, Section B* **74B**, pp. 47-54.

Robinson, S.M. (1972). Extension of Newton's method to nonlinear functions with values in a cone, *Numerische Mathematik*, **19**, pp. 341-347.

Robinson, S.M. (1973). Computable error bounds for nonlinear programming, *Mathematical Programming*, **5**, pp. 235-242.

Rockafellar, R.T. (1981). *The Theory of Subgradients and its Application to Problems of Optimization: Convex and Nonconvex Functions*, Heldermann-Verlag, Berlin.

Rohn, J. (1981). Strong solvability of interval linear programming problems, *Computing*, **26**, pp. 79-82.

Rokne, J. (1985). A low complexity rational centered form, *Computing*, **34**, pp. 261-263.

Rokne, J. (1986). Low complexity k-dimensional centered forms, *Computing*, **37**, pp. 247-253.

Rokne, J. Bao, P. (1987). Interval Taylor forms, *Computing*, **39**, pp. 247-259.

Rokne, J. Bao, P. Low Complexity k-dimensional Taylor Forms, *Applied Math. and Comp.*, to appear.

Rubinstein, Y., Weissman, I. (1979). The Monte Carlo method for global optimization, *Can. Cent. Etud. Rech. Oper.*, **21**, pp. 143-147.

Schittkowski, K. (1979). An adaptive precision method for nonlinear optimization problems, *SIAM Journal on Control. Optimization*, **17**, pp. 87-98.

Schittkowski, K. (ed.) (1985). *Computational Mathematical Programming*, Springer-Verlag, Berlin.

Schnabel, R.B. (1982). Determining feasibility of a set of nonlinear inequality constraints, *Mathematical Programming Study*, **16**, pp. 137-148.

Schnabel, R.B. (1983). Conic methods for unconstrained minimization and tensor methods for nonlinear equations. In: Bachem-Grötschel-Korte (eds.) (1983), pp. 417-438.

Schnabel, R.B., Frank, P.D. (1984). Tensor methods for nonlinear equations, *SIAM Journal on Numerical Analysis*, **21**, pp. 815-843.

Schnabel, R.B., Frank, P.D. (1987). Solving systems of nonlinear equations by tensor methods. In: Iserles-Powell (eds.) (1987), pp. 245-272.

Schnabel, R.B., Koontz, J.E. (1985). A modular system of algorithms for unconstrained minimization, *ACM Trans. on Math. Software*, **11**, pp. 419-440.

Schoen, F. (1982). On a sequental search strategy in global optimization problems, *Calcolo*, **19**, pp. 321-333.

Schrempp, G. (1984). Intervall-Iterationsverfahren für nichtlineare Gleichungssysteme: Theorie und praktischer Vergleich am Computer. *Freiburger Intervall-Berichte* 84/6 Institut für Angewandte Mathematik, Universität Freiburg.

Schwandt, H. (1984). A symmetric iterative interval method for systems of non-linear equations, *Computing*, **33**, pp. 153-154.

Sengupta, S. (1981). Global nonlinear optimization, Ph.D. thesis, Washington State University, Pullman, Washington.

Shearer, J.M., Wolfe, M.A. (1985). Some computable existence, uniqueness, and convergence tests for nonlinear systems, *SIAM Journal on Numerical Analysis*, **22**, pp. 1200-1207.

Shearer, J.M., Wolfe, M.A. (1986). A note on the algorithm of Alefeld and Platzöder for systems of nonlinear equations, *SIAM J. Sci. Stat. Comp.*, **7**, pp. 362-369.

Shen, Z., Zhu, Y. (1985). An interval version of Shubert's iterative method for the localization of the global maximum, *Freiburger Intervall-Berichte* 85/7, Institut für Angewandte Mathematik, Universität Freiburg, pp. 37-47.

Shohdohji, T. (1977). An algorithm for obtaining global optima for multi-variable multi-modal functions, *J. Oper. Res. Soc. Jpn.*, **20**, pp. 311-320.

Shor, N.Z. (1983). Generalized gradient methods of nondifferentiable optimization employing space dilatation operations. In: Bachem-Grötschel-Korte (eds.) (1983), pp. 501-529.

Shubert, R.O. (1972). A sequential method seeking the global maximum of a function, *SIAM Journal on Numerical Analysis*, **9**, pp. 379-388.

Shusterman, L.B. (1979). Successive elimination in the search for a global optimum of multiextremal algebraic functions, *Izv. Vuz. Radioelektron (USSR)*, **22**, pp. 58-63.

Simmons, D.M. (1975). *Nonlinear Programming for Operations Research*, Prentice-Hall, Englewood Cliffs, New Jersey.

Singh, M.G., Hassan, M.F. (1977). Local and global optimal control for non-linear systems using two-level methods, *Int. J. Syst. Sci.*, **8**, pp. 1375-1383.

Sisser, F.S. (1981). Elimination of bounds in optimization problems by transforming variables, *Mathematical Programming*, **20**, pp. 110-121.

Skelboe, S. (1974). Computation of rational interval functions, *BIT*, **14**, pp. 87-95.

Spircu, L. (1979). Cluster analysis in global optimization, *Econ. Comput. and Econ. Cybern. Stud. and Res.*, **4**, pp. 43-50.

Stewart, G.W. (1973). *Introduction to Matrix Computations*, Academic Press, New York.

Stewart, N. (1973). Interval arithmetic for guaranteed bounds in linear programming, *J. Optim. Theor. and Appl.*, **12**, pp. 1-5.

Stoer J., Witzgall, C. (1970). *Convexity and Optimization in Finite Dimensions I*, Springer-Verlag, Heidelberg.

Strongin, R.G. (1973). On the convergence of an algorithm for finding a global extremum, *Eng. Cybern.*, **11**, pp. 549-555.

Sunaga, T. (1958). Theory of an interval algebra and its application to numerical analysis, *RAAG Memoirs*, **2**, pp. 547-564.

Tapia, R. (1987). Research in numerical optimization. In: Moore (ed.) (1988).

Timonov, L.N. (1977). An algorithm for search of a global extremum, Translated in *Eng. Cybern.*, **15**, pp. 38-44.

Torn, A.A. (1976). Cluster analysis using seed points and density-determined hyperplanes with an application to global optimization, IEEE *et al*, 3rd International Joint Conference on Pattern Recognition, New York: *IEEE*, New York, pp. 394-B.

Tuy, H.A. (1985). A general deterministic approach to global optimization via d.c. programming. In: J.-Hiriart-Urruty (eds.), *FERMAT Days 1985: Mathematics for Optimization*, Elsevier Sci. Publishers, pp. 273-303.

Vilkov, A., Zhidkov, N., Shchedrin, B. (1975). A method of finding the global minimum of a function of one variable, *USSR Comp. Math. and Math. Phys.*, **15**, pp. 221-223.

Walster, G.W. (1987). Philosophy and practicalities of interval arithmetic. In: Moore (ed.) (1988).

Walster, G.W., Hansen, E.R., Sengupta, S. (1985). Test results for a global optimization algorithm. In: Boggs-Byrd-Schnabel (eds.) (1985), pp. 272-287.

Wang, B.C., Luus, R. (1978). Reliability of optimization procedures for obtaining global optimum, *AICHE J.*, **24**, pp. 619-626.

Warmus, M. (1956). Calculus of approximations, *Bull. Acad. Polon. Sci. Cl. III*, **4**, pp. 253-259.

Wengert, R.E. (1964). A simple automatic derivative evaluation program, *Comm. ACM*, **7**, pp. 463-464.

Westerberg, A.W., Shah, J.V. (1978). Assuring a global optimum by the use of an upper bound on the lower (dual) bound, *Comput. and Chem. Eng.*, **2**, pp. 83-92.

Williams, H. P. (1986). Fourier's method of linear programming and its dual, *Amer. Math. Monthly*, **93**, pp. 681-694.

Wolfe, M.A. (1978). *Numerical Methods for Unconstrained Optimization*, Van Nostrand Reinhold Company, New York.

Wolfe, M.A. (1980). A modification of Krawczyk's algorithm, *SIAM Journal on Numerical Analysis*, **17**, pp. 376-379.

Wolfe, P. (1975). A method of conjugate subgradients for minimizing nondifferentiable functions, *Mathematical Programming Study*, **3**, pp. 145-173.

Wood, G.R. (1975). Multidimensional bisection and global minimization, University of Canterbury, Preprint.

Yohe, J.M. (1979). Software for interval arithmetic: a reasonably portable package, *ACM Trans. on Math. Software*, **5**, pp. 50-63.

Yohe, J.M. (1980). Interval analysis comes of age, *SIAM News*, **13**, pp. 1,8.

Zaliznyak, N.F., Ligun, A.A. (1978). Optimal strategies for seeking the global maximum of a function, *Zh. vychisl. Mat. mat. Fiz. (USSR)*, **182**, pp. 314-321.

Zangwill, W.I. (1969). *Nonlinear Programming*, Prentice-Hall, Englewood Cliffs, NJ.

Zhulenev, S.V. (1974). On a method for seeking a global extremum, *Eng. Cybern. (trans.)*, **12**, pp. 32-38.

Zimmermann, K.A. (1979). Generalization of convex functions, *Ekon.-mat Obz.*, **15**, pp. 147-158.

Zowe, J. (1985). Nondifferentiable optimization. In: Schittkowski, K. (ed.) (1985), pp. 321-356.

Zwart, P.B. (1973). Nonlinear programming: counter-examples to two global optimization algorithms, *Oper. Res.*, **21**, pp. 1260-1266.

Notation

$\mathbf{R}$	Set of reals, 10, 26
$\mathbf{N}$	Set of non-negative integers, 97
$\mathbf{R}^{k \times l}$	Set of $k \times l$-matrices over $\mathbf{R}$, 32
$\mathbf{R}^k = \mathbf{R}^{k \times 1}$	Set of k-vectors (column vectors) over $\mathbf{R}$, 11, 32
$\mathbf{I}$	Set of real compact intervals, that is, $[a, b]$, etc., 27
$\mathbf{I}^{k \times l}$	Set of $k \times l$ - matrices over $\mathbf{I}$, 32
$\mathbf{I}^k = \mathbf{I}^{k \times 1}$	Set of k-vectors (column vectors) over $\mathbf{I}$, 32
$\mathbf{I}_\infty$	Set of all closed intervals (bounded or not), that is, $[a, b]$ (bounded), $[a, \infty), (-\infty, \infty)$ (unbounded), etc., 137
$\overline{\mathbf{R}} = \mathbf{R} \cup \{-\infty, \infty\}$	The two-point compactified real line, 136
$\overline{\mathbf{I}}$	Set of all compact intervals over $\overline{\mathbf{R}}$, for example, $[a, \infty], \infty = [\infty, \infty], \overline{\mathbf{R}} = [-\infty, \infty]$, etc., 136
$\overline{\mathbf{R}^m} = \overline{\mathbf{R}}^m$	The m-fold topological product of $\overline{\mathbf{R}}$, 136

$$\left.\begin{array}{rl}\mathbf{I}(D) &= \{Y \in \mathbf{I}^m : Y \subseteq D\} \\ \bar{\mathbf{I}}(D) &= \{Y \in \bar{\mathbf{I}}^m : Y \subseteq D\} \\ \mathbf{I}_\infty(D) &= \{Y \in \mathbf{I}_\infty^m : Y \subseteq D\}\end{array}\right\} \text{ for } D \subseteq \overline{\mathbf{R}^m}, \ 32, \ 137$$

x_n, X_n	Denote both iterates as well as components of $x = (x_1, \ldots, x_m)^T$ or $X = (X_1, \ldots, X_m)^T$, 11
x^T, X^T	Transpose of x or X, 10, 32
x^i	Denotes the components of x in exceptional cases when misunderstandings with iterates can occur, 137
$\mathbf{C}^n$	Class of functions which have a n-th continuous derivative or partial derivatives. $\mathbf{C}^1$-functions are also called smooth functions, 17
$w(A)$	Width of intervals and boxes, 25, 31, 32
mid$A$	Midpoint of intervals and boxes, 25, 32
lb$A$	Left endpoint (lower boundary) of an interval $A \in \bar{\mathbf{I}}$, 31
ub$A$	Right endpoint (upper boundary) of an interval $A \in \bar{\mathbf{I}}$, 31
lb$A$	Vector of left endpoints of a box $A \in \mathbf{I}^m$, 32
ub$A$	Vector of right endpoints of a box $A \in \mathbf{I}^m$, 32
$0 < A$	$0 < \mathrm{lb}A_i, i = 1, \ldots, m$ for $A \in \mathbf{I}^m$, 33

Notation

$0 \leq A$	$0 \leq \text{lb} A_i, i = 1, \ldots, m$ for $A \in \mathbf{I}^m$, etc., 33
$A < B$	$\text{ub} A_i < \text{lb} B_i, i = 1, \ldots, m$ for $A, B \in \mathbf{I}^m$, 33
$A \leq B$	$\text{ub} A_i \leq \text{lb} B_i, i = 1, \ldots, m$ for $A, B \in \mathbf{I}^m$, 33
$x \vee y$	Smallest interval vector which contains $x, y \in \mathbf{R}^m$, 34
$X \vee Y$	Smallest interval vector (or interval matrix) which contains $X, Y \in \mathbf{I}^m$ (or $X, Y \in \mathbf{I}^{m \times m}$), 34, 91
$f : \mathbf{R}^m \to \mathbf{R}$	Mainly the objective function (function to be minimized), 10
x^*	Any global minimizer, 11
X^*	Set of all global minimizers, 73
f^*	Global minimum, 11, 73
$f'(x) \in \mathbf{R}^{m \times 1}$	Gradient of f at x $$f'(x) = \left(\frac{\partial f(x)}{\partial x_1}, \ldots, \frac{\partial f(x)}{\partial x_m}\right)^T, \ 12$$
$f''(x) \in \mathbf{R}^{m \times m}$	Hessian matrix of f at x $$f''(x) = \left(\frac{\partial^2 f(x)}{\partial x_i \partial x_j}\right)_{\substack{i=1,\ldots,m \\ j=1,\ldots,m}}, \ 17$$

$\partial f(x)$	Generalized gradient of f at x $$\partial f(x) = ((\partial f)_1(x), \ldots, (\partial f)_m(x))^T, \ 44$$
$g_i(x) \leq 0$	Inequality constraints, $i = 1, \ldots, k$, 10
$g : \mathbf{R}^m \to \mathbf{R}^k$	Inequality constraint functions (vector notation), 10
$J_g(x)$	$= (g'_1(x), \ldots, g'_k(x))^T$ Jacobian matrix of g at x, 12
$h_i(x) = 0$	Equality constraints, $i = k+1, \ldots, r$, 10
$h : \mathbf{R}^m \to \mathbf{R}^{r-k}$	Equality constraint functions (vector notation), 10
$J_h(x)$	$= (h'_{k+1}(x), \ldots, h'_r(x))^T$ Jacobian matrix of h at x, 12
$\Box f(Y)$	Range of f over Y, 34, 35
$f(Y)$	Natural interval extension of f at Y, 36
$Y_n \to A$	Convergence of Y_n to A with respect to the Hausdorff-metric of compact sets, 79
$d(A, B)$	Hausdorff-metric of compact sets, 78, 140
I	Identity matrix, 19, 56
L	List, 63, 77, 78
$E \pm k$	$10 \pm k$, 66, 90

Index

accumulation point of an interval sequence 78
active index set 13
active inequality constraint 11
automatic differentiation 42, 120
bisection strategy 75
bounding principle 77
box 32, 73
branch and bound principle 74, 118
branching principle 77
compactified problem 138
compactified unbounded intervals 135
conjugate gradient methods 21 ff
constraint qualification 13
constraints 10
convergence order 38
convergence order of a set of sequences 99
converges arbitrarily slow 99
convex-simplex problem 191
direct search methods 22
discrete optimization problem 11
distributive law 28
equality constraint functions 161
equality constraints 161
exact penalty function 16

excess-width 38
exhaustion principle 118
factorable programming 120
feasible area 161, 162
feasible point 11, 161
feasible region 11, 161
feasible set 11, 161
Fletcher-Reeves method 192
fundamental property of interval arithmetic 36
generalized gradient 44
generalized Lagrangian function 13
global constrained minimization problem 162
global constrained optimization problem 161
global maximizer 74
global maximum 74
global minimizer 11, 73, 162
global minimum 11, 73, 162
global minimum point 11, 73, 162
global minimum value 11
global unconstrained minimization problem 73
global unconstrained optimization problem 73
global unconstrained optimiza-

227

tion problem over unbounded domains 138
Hansen's algorithm 110 *ff*
Hansen-Greenberg realization 53 *ff*
Hausdorff-metric 78
Ichida-Fujii algorithm 108 *ff*
inactive inequality constraint 11
inclusion 33
inclusion function 35
inclusion isotonicity 29, 34
inclusion principle of interval arithmetic 27
inclusion principle of machine interval arithmetic 31
indeterminate 165
inequality constraint functions 161
inequality constraints 161
infeasible area 161
infeasible point 161
integer optimization problem 11
interval arithmetic operations 27 *ff*
interval matrix operations 33
interval Newton algorithm 52 *ff*
interval Newton algorithm, one iteration of 53
interval variable 34
interval vector operations 33
John criterion 12
Kuhn-Tucker conditions 13
Lagrangian function 14
Lagrangian multiplier 14
leading box 78
leading pair 78
linear interval equation 52

linear optimization problem 10
Lipschitz constant 38
Lipschitz inclusion function 38
local maximum 12
local maximum point 12
local minimizer 11, 162
local minimum 11
local minimum point 11, 162
local minimum value 11
local opotimization 1
local optimum 12
machine-finite intervals 152
machine-infinite intervals 152
machine interval arithmetic 30 *ff*
machine intervals 30, 136
mathematical programming problem 10
meanvalue form 39
meanvalue form function 39
midpoint 32, 139
midpoint test 108
monotonicity test 44, 122, 146 *ff*
Moore's existence test 170
Moore-Skelboe algorithm 77
natural interval extension 36, 151
Newton method 194
Newton's method 17
Newton's method for unconstrained optimization 19
noncompactified unbounded intervals 136
nonlinear programming problem 10
non-wasteful 138
objective function 10, 161

INDEX

optimality conditions 12 ff
optimization problem 1, 10
order 38
penalty factor 16
penalty function 191
point intervals 27
preprocessing exhaustion algorithm 186
quadratic optimization problem 11
quasi Newton methods 20 ff
range 35
recursive differentiation 120
regularity condition 13
relaxation parameter 174
Robinson-Daniel method 191
secant methods 21
slopes 41
steepest descent algorithm 21
steepest descent methods 21 ff
subdistributive law 28
subdivision strategy 75
symmetric interval 30
Taylor form 42
Taylor form function 42
unconstrained minimization 16 ff
unconstrained minimization problem 16
unconstrained problem 10
variable metric methods 20
width 32, 139

Mathematics and its Applications

Series Editor: G. M. BELL, Professor of Mathematics, King's College London (KQC), University of London

Faux, I.D. & Pratt, M.J.	Computational Geometry for Design and Manufacture
Firby, P.A. & Gardiner, C.F.	Surface Topology
Gardiner, C.F.	Modern Algebra
Gardiner, C.F.	Algebraic Structures: with Applications
Gasson, P.C.	Geometry of Spatial Forms
Goodbody, A.M.	Cartesian Tensors
Goult, R.J.	Applied Linear Algebra
Graham, A.	Kronecker Products and Matrix Calculus: with Applications
Graham, A.	Matrix Theory and Applications for Engineers and Mathematicians
Graham, A.	Nonnegative Matrices and Applicable Topics in Linear Algebra
Griffel, D.H.	Applied Functional Analysis
Griffel, D.H.	Linear Algebra
Guest, P. B.	The Laplace Transform and Applications
Hanyga, A.	Mathematical Theory of Non-linear Elasticity
Harris, D.J.	Mathematics for Business, Management and Economics
Hart, D. & Croft, A.	Modelling with Projectiles
Hoskins, R.F.	Generalised Functions
Hoskins, R.F.	Standard and Non-standard Analysis
Hunter, S.C.	Mechanics of Continuous Media, 2nd (Revised) Edition
Huntley, I. & Johnson, R.M.	Linear and Nonlinear Differential Equations
Jaswon, M.A. & Rose, M.A.	Crystal Symmetry: The Theory of Colour Crystallography
Johnson, R.M.	Theory and Applications of Linear Differential and Difference Equations
Johnson, R.M.	Calculus: Theory and Applications in Technology and the Physical and Life Sciences
Jones, R.H. & Steele, N.C.	Mathematics of Communication
Jordan, D.	Geometric Topology
Kelly, J.C.	Abstract Algebra
Kim, K.H. & Roush, F.W.	Applied Abstract Algebra
Kim, K.H. & Roush, F.W.	Team Theory
Kosinski, W.	Field Singularities and Wave Analysis in Continuum Mechanics
Krishnamurthy, V.	Combinatorics: Theory and Applications
Lindfield, G. & Penny, J.E.T.	Microcomputers in Numerical Analysis
Livesley, K.	Engineering Mathematics
Lord, E.A. & Wilson, C.B.	The Mathematical Description of Shape and Form
Malik, M., Riznichenko, G.Y. & Rubin, A.B.	Biological Electron Transport Processes and their Computer Simulation
Massey, B.S.	Measures in Science and Engineering
Meek, B.L. & Fairthorne, S.	Using Computers
Menell, A. & Bazin, M.	Mathematics for the Biological Sciences
Mikolas, M.	Real Functions and Orthogonal Series
Moore, R.	Computational Functional Analysis
Murphy, J.A., Ridout, D. & McShane, B.	Computation in Numerical Analysis
Nonweiler, T.R.F.	Computational Mathematics: An Introduction to Numerical Approximation
Ogden, R.W.	Non-linear Elastic Deformations
Oldknow, A.	Microcomputers in Geometry
Oldknow, A. & Smith, D.	Learning Mathematics with Micros
O'Neill, M.E. & Chorlton, F.	Ideal and Incompressible Fluid Dynamics
O'Neill, M.E. & Chorlton, F.	Viscous and Compressible Fluid Dynamics
Page, S. G.	Mathematics: A Second Start
Porter, T. & Cordier, J.	Model Formulation Analysis
Prior, D. & Moscardini, A.O.	Model Formulation Analysis
Rankin, R.A.	Modular Forms
Scorer, R.S.	Environmental Aerodynamics
Smith, D.K.	Network Optimisation Practice: A Computational Guide
Shivamoggi, B.K.	Stability of Parallel Gas Flows
Stirling, D.S.G.	Mathematical Analysis
Sweet, M.V.	Algebra, Geometry and Trigonometry in Science, Engineering and Mathematics
Temperley, H.N.V.	Graph Theory and Applications
Thom, R.	Mathematical Models of Morphogenesis
Thurston, E.	Primary Mathematics: Teaching and Learning
Townend, M. S.	Mathematics in Sport
Towend, M.S. & Pountney, D.C.	Computer-aided Engineering Mathematics
Twizell, E.H.	Computational Methods for Partial Differential Equations
Twizell, E.H.	Numerical Methods, with Applications in the Biomedical Sciences
Vince, A. and Morris, C.	Mathematics for Computer Studies
Walton, K., Marshall, J., Gorecki, H. & Korytowski, A.	Control Theory for Time Delay Systems
Warren, M.D.	Flow Modelling in Industrial Processes
Wheeler, R.F.	Rethinking Mathematical Concepts
Willmore, T.J.	Total Curvature in Riemannian Geometry
Willmore, T.J. & Hitchin, N.	Global Riemannian Geometry

Numerical Analysis, Statistics and Operational Research
Editor: B. W. CONOLLY, Professor of Mathematics (Operational Research), Queen Mary College, University of London

Beaumont, G.P.	Introductory Applied Probability
Beaumont, G.P.	Probability and Random Variables
Conolly, B.W.	Techniques in Operational Research: Vol. 1, Queueing Systems
Conolly, B.W.	Techniques in Operational Research: Vol. 2, Models, Search, Randomization
Conolly, B.W.	Lecture Notes in Queueing Systems
Conolly, B.W. & Pierce, J.G.	Information Mechanics: Transformation of Information in Management, Command, Control and Communication
French, S.	Sequencing and Scheduling: Mathematics of the Job Shop
French, S.	Decision Theory: An Introduction to the Mathematics of Rationality
Griffiths, P. & Hill, I.D.	Applied Statistics Algorithms
Hartley, R.	Linear and Non-linear Programming
Jolliffe, F.R.	Survey Design and Analysis
Jones, A.J.	Game Theory
Kapadia, R. & Andersson, G.	Statistics Explained: Basic Concepts and Methods
Moscardini, A.O. & Robson, E.H.	Mathematical Modelling for Information Technology
Moshier, S.	Mathematical Functions for Computers
Oliveira-Pinto, F.	Simulation Concepts in Mathematical Modelling
Ratschek, J. & Rokne, J.	New Computer Methods for Global Optimization
Schendel, U.	Introduction to Numerical Methods for Parallel Computers
Schendel, U.	Sparse Matrices
Sehmi, N.S.	Large Order Structural Eigenanalysis Techniques: Algorithms for Finite Element Systems
Späth, H.	Mathematical Software for Linear Regression
Spedicato, E. and Abaffy, J.	ABS Projection Algorithms
Stoodley, K.D.C.	Applied and Computational Statistics: A First Course
Stoodley, K.D.C., Lewis, T. & Stainton, C.L.S.	Applied Statistical Techniques
Thomas, L.C.	Games, Theory and Applications
Whitehead, J.R.	The Design and Analysis of Sequential Clinical Trials

ELECTRONIC IMAGING
MARKETING RESEARCH INFORMATION CENTER
THE DUPONT COMPANY